Praise for *In*

"Addressing the threat of global warming and getting all of humanity to live with one another and the natural world in harmony is very possible. In this book, Wilford shows us the way."

Desmond Tutu, Nobel Peace Prize Laureate

"*In Our Hands* is a brilliant perspective from one of the world's most authentic authorities on social entrepreneurship and climate change. Wilford Welch hands us the missing link as to how we can reach across the generational gap between millennials and boomers and evolve together the solutions to the disastrous climate change predictions. He invites us into a view of our world from 2050—an ingenious way to think, and subsequently act, outside the proverbial 'box.' This is a 'must read' for any age interested in our survival as a species."

**Diane V. Cirincione, Ph.D.,
Attitudinal Healing International**

"I have a long history with this subject matter, and this is an impressive book for many reasons: It is concise, accurate, readable, and even well-written! I don't take any of these attributes for granted when reviewing a book, so Wilford's book was refreshing for all these reasons."

Jeff Battis, Book Passage Bookstore, San Francisco

"Wilford Welch has written a fabulous, accessible, brilliant, and critical book that empowers each of us to take action to reverse global warming and create the world we want. This is an easy read, and so inspiring and motivating that you will be in immediate action as a result. Everyone alive should have this book in their hands! I loved every word. Buy it, read it, use it, share it!"

Lynne Twist, author, *The Soul of Money*

"LOVED THIS BOOK! I read it a couple of times and think you have done an AMAZING JOB—causes, solutions, actions, resources. I especially like the way you start at 2050 and see the actions that were taken that made the difference; and I love all of the references you give."

Sharon Roe, Ph.D., Philosophy and Religion

"This is the quintessential book on climate change! Welch's passionate and practical easy-to-read book is filled with compelling scenarios and resources to move everyone to action. You get to choose your adventure: the probable future or the possible future! Choose this book, and buy more to give to your friends. I did, and I am part of the solution!"

Barbara Meyer, Houston, Texas

"Wilford Welch's guide for intergenerational actions to solve the climate crisis is exactly what we need, exactly when we need it. As the baby boomer generation ages and we prepare for a $50 trillion intergenerational wealth transfer between boomers and millennials, it is essential that we bridge the age gap and begin to work together to create a more just and regenerative world for everyone. I can't wait to share this book with friends and family of all ages and to work on one of the most pressing issues of our time."

Eli Utne, millennial teacher of sustainability practices to underserved youth

"*In Our Hands* is a very important book to help galvanize those who say they are concerned but are not really engaged to wake up and take action. It's both frightening and sobering. Also, the back section of the book—the work area and resources—is very important and an effective call to action. The book is well written and masterfully edited. None of this is far-fetched. It's happening now and will only worsen. There are no borders and impenetrable barriers to be erected to combat a potential extinction of life."

Gordon Haight, international publisher and marketing consultant

"This new book, *In Our Hands,* provides a succinct and effective overview of the issues involved in global warming and then offers a pathway to success in solving the climate crisis. It is also an accessible and valuable handbook—a guide to who is doing what and how you can get involved in the climate movement. This small book makes a big contribution to galvanizing the actions that are needed now."

Bill Twist, CEO, Pachamama Alliance

"This awesome book presents a fresh intergenerational perspective on how to approach the climate crisis. It presents both the probable future that we are headed toward and the positive future if we take action, as well as the actions that we must take to make it possible."

David B. Room, millennial activist

"The scenarios are inspired and thought-provoking. Clearly, you care deeply, profoundly, and that comes across in every page."

**Susan Collin Marks, peace ambassador,
Search for Common Ground**

"This slim volume is the perfect place to start tackling the greatest problem of our time. *In Our Hands* makes crystal clear why all citizens of planet Earth must immediately join the fight against man-made climate change. But this essential handbook goes way beyond mere motivation by also detailing the simple but powerful steps we all can take right now, and by cataloging the resources we can use to do so. Essential reading!"

V. Ricci, author, *Clueless*

"This concise, well-written look on the climate crisis will leave you with hope. Welch does so by suggesting positive actions we can all take to fight climate change and create a better world."

Will Parrinello, documentary filmmaker

"There is not a single existential problem facing the United States, or the entire world for that matter, that does not have one or more attainable and affordable solutions provided that there is the proper leadership and the collective will to solve them. Wilford Welch has gone a long way in this important contribution to our collective knowledge about our future by outlining the problems we face caused by our contribution to global warming and what must be done to address them before it is too late."

Peter Bergh, landscape architect

IN OUR HANDS

A HANDBOOK FOR INTERGENERATIONAL ACTIONS TO SOLVE THE CLIMATE CRISIS

Be the change

Blessings

Wilford Welch

Wilford H. Welch

ISBN for the print edition: 978-0-9988879-0-6
ISBN for the e-book edition: 978-0-9988879-1-3

Dedication

To
Gordon Gund and Jerry Jampolsky
Two dear friends
Both now blind
Both at one time could see perfectly
Now, it seems they can see humanity's strengths and
weaknesses far more clearly

And, to the next generations
Ashley, John, Shandy, Hans, Charlie, Teresa
Noelle, Aydan, Finley, Ayla, Ashley, Kennedy, Sophia
and
Coralie

May we all work together to ensure that there will be
many generations to come

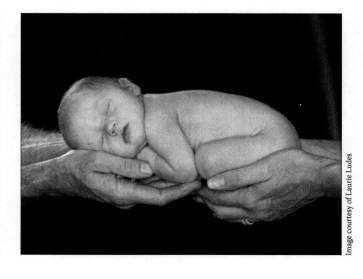

Image courtesy of Laurie Ludes

Contents

➢ Global warming is indeed happening, and it is primarily caused by human activities.

➢ We can get the global-warming crisis under control by 2050.

➢ "We" means all of us doing our part – in our personal, community, and professional lives.

➢ These actions do not have to be difficult. They will add pleasure and value to your life as well as to the lives of your children and grandchildren.

➢ If we do not rise to meet this challenge, mother nature will remind us who is ultimately in charge.

This book has been written with these convictions at its core – and with great hope for our common future.

Preface

My awakening to the possibility that global warming, over-population, and overdevelopment could lead to the collapse of humanity occurred two years ago in Southern Ethiopia. My wife Carole and I were leading a trip there, exploring how the tribes were dealing with the many natural and human forces impacting their world. The Hamar people are the descendants of the very first humans who lived on our planet some 2.8 million years ago. For centuries they have lived in the fertile Omo Valley raising cattle and crops.

Author with Hamar villagers in Southern Ethiopia.

Kala, our twenty-nine year old guide, told us that he feared for the future of many of the tribes who depend on continued access to their pastoral homelands. He said the weather was getting hotter and hotter, and the rains were no longer predictable. He noted that one can no longer drive on the roads without seeing livestock that have died from lack of food and water. He said that much of the river water the tribes had always relied on was being diverted by the government to the large plots of land the government had taken from the tribes and leased to investors from Turkey, Saudi Arabia, and China to grow crops for export. Kala was sure that further marginalization and even starvation lay ahead for his tribe and would lead to clashes with their national government, which seemed more interested in growing the national economy than ensuring the well-being of his people.

Seeking water in Southern Ethiopia.

And sure enough, a few months after we left, tribal frustrations boiled over when government soldiers put a shooting range right next to one of the villages, and the tribesmen retaliated by killing twelve soldiers. More recently, government forces killed more than six hundred tribes-men over a period of several months. Kala was picked up while driving one night, interrogated, and held in jail for three days on unspecified charges. Clear that his life was in danger due to repressive government tactics, he fled to a nearby country where he went into hiding, desperately wanting to see his wife and children again and get on with building his business, but fearing for his life if he returned.

My experience with Kala and his tribe in Southern Ethio-pia brought into focus what I had long been concerned about but that had not moved me to take action against global warming. It was clear that what I was witnessing in Ethiopia were the same forces that could lead to the extinc-tion of humanity if we did not change our ways. Those forces include: too many people competing for dwindling amounts of fresh water and arable land; people's health and livelihoods being ravaged by global warming; and gov-ernments determined to achieve economic growth "at all costs" – by which I mean the heavy costs to both people and the planetary ecosystem that we rely on to support us.

As I write this, Kala's story continues to unfold. He may someday be able to return to his family, home, and busi-ness, but none of the conditions that are threatening him, his country, and his people have changed. I realized that beyond being a friend and supporter of Kala, I wanted to bring whatever experience and resources I had available to focus attention on the role global warming, popula-tion growth, and overdevelopment are having not only on Ethiopia but on the whole world. Thus, this small book – a labor of love that is coming from a deep concern for our common future.

Introduction

Global warming and climate change do not have to be scary topics, and this book seeks to show how you can turn the fear you may have into an opportunity, not only to solve the climate challenge, but to create a better world for your children and grandchildren.

To achieve this positive future requires that we shift some of our focus away from the busyness of our daily lives and toward the climate challenge, which to many readers may also feel like a distant concern you can do little about. But a distant concern it is not, and scary it does not have to be. This book shows just what is possible if, individually and collectively, we focus our attention and take some of the relatively easy actions noted in Chapter 4 and Chapter 5.

According to research by the Yale Program on Climate Change Communication, 18 percent of Americans are alarmed by climate change and are taking individual, consumer, and political actions to address it, while another 34 percent say they are concerned but in reality are not engaged. These 34 percent come from all segments of society: young and old, rich and poor, from red and blue states, and from most professions. They are sitting on the sidelines hoping that technological breakthroughs, government initiatives, and/or the efforts of those 18 percent will solve the problem. But that 18 percent is not enough to win this race.

This book seeks to inspire many of those 34 percent "concerned-but-not-engaged citizens" to jump in the boat and start rowing. If they do, the movement will become unstoppable.

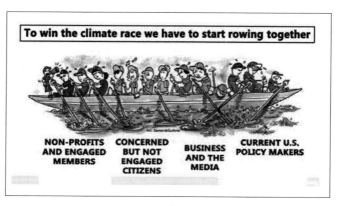

The primary audience for this book is indeed this 34 percent of Americans, because if the United States does not rise to meet this challenge it will pull the rest of the world down with it. The actions outlined in this book are however just as applicable to all citizens of the United States and all other societies of our beautiful world.

If we are going to effectively solve global warming and climate-change challenges, we must also be attentive to their root causes, namely the massive growth in population, consumption, and development that has become the norm since the end of the Second World War. And we must address the many effects of global warming and climate change, among them insufficient fresh water and arable soil to support so many people, which is resulting in millions of climate refugees. Finally, because most of our actions come from the values we hold, this book also explores those values embedded in our cultures that have contributed to this crisis and will need to be modified if all 7.5 billion of us are to achieve a sustainable future.

These are the many interrelated issues discussed in this book, along with the actions the average person can take, individually and collectively, to address many of them. There is plenty of work to go around, much of it very rewarding.

Thousands of governmental, non-governmental organizations (NGOs), and nonprofits in the United States and around the world are doing extraordinary work informing the public about these issues and coming up with solutions, but the current citizen-based movement is not massive enough or focused enough to win the race. I hope this book and the resources it provides will be a contribution to the U.S. and worldwide mobilization that is now needed as a matter of urgency, and that you, the reader, will not only keep reading but will also join the movement and become part of the solution.

What concerns me

Before discussing all the things that give me great hope for the future, I would like to share some concerns.

I am concerned that any progress that has been made to address climate change will be derailed by politicians who have different political interests. An example of this is President Donald Trump's withdrawal of the United States from the Paris climate accord. Many of our political leaders don't believe in climate change; others aren't motivated to do anything about it. Several political leaders stand firmly with the fossil-fuel industry in its commitment to stimulate more oil, coal, and natural gas production and consumption – and thus more carbon pollution. Their support for an economic growth-at-all-costs agenda may seem logical – as well as politically advantageous – over the near term. But if history proves these politicians wrong, as I believe will be the case, they will have won a near-term political victory but lost the war — the biggest war humanity has ever lost.

It is also discouraging to witness the breakdown of the U.S. policy-making process that has occurred in recent years because of bitter political partisanship. During my years as a U.S. diplomat in Washington, D.C., and Asia during the Johnson and Nixon administrations, politics was often a rough and tumble affair, but at the end of the day the focus was on making good policy decisions. Senators would vigorously debate each other during the day, but might then go out for dinner together. Compromise was seen as necessary on the road to good policy. Today it appears that partisan politics trumps good policy, science is ignored if it does not support one's political ideology and compromise is seen as tantamount to treason.

Another concern is that governments, as well as individuals, have a propensity to ignore or deny "inconvenient truths," including all the signs of global climate change that are already evident both from the science and from the events taking place every day around the world. If an event destroys the life or livelihood of someone we do not know or live near, we tend to ignore or discount it. Thus, I fear it may take a truly catastrophic event that personally affects us all to move us to take concerted, collective action. I had hoped that Hurricane Katrina, which crippled New Orleans, Hurrican Sandy, which damaged so much of the Northeast, Hurrican Harvey, which damaged so much of Houston and other areas in Texas, Hurrican Irma, which destroyed many islands in the Carribbean and damaged South Florida, and Hurricane Maria, which crippled Puerto Rico, along with the recent heat waves and devastating forest fires in the western United States, would have been enough to focus many more of us on the reality of global warming and climate change. My hope of course is that we take action before such a catastrophe occurs – and thus the reason for this book. I see it as a handbook for a "rEvolution" – a revolution that changes the values we hold and that motivates the actions we take.

Santa Rosa, California, October 2017. Some of the 8,400 structures destroyed by a devastating wildfire in this area. The effects of global warming are linked to wildfires.

Image courtesy of Wilford Welch

You probably share my concern that humans may not have the capacity to drop our animosities toward one another as individuals or political parties and focus on the "we" as much as the "me." Awakening to the need to come together to save our common home is now essential. We rose to the occasion on December 8, 1941, in response to Japanese and Nazi challenges, and I believe we can rise to the occasion again to address the current existential threat.

Finally, I am concerned that we will assume that technological advances alone will solve the climate challenges we are facing. While technology is critically important and progressing rapidly, especially in the development of renewable energy sources, I am clear that addressing the climate crisis will also require us all to make substantial changes in both our values and behaviors in a very short period of time. This too is a major challenge.

What gives me hope

Despite all the concerns just expressed, there are many reasons to be hopeful. For one, even though it is rarely reported in the media, amazing work has been and is being done by millions of individuals and hundreds of thousands of NGOs around the world to bring about the changes highlighted in the pages that follow. They have laid the groundwork for the "rEvolution," and it is now time for the rest of us to move their hard work and wisdom forward and put it into action.

I am convinced that the millennials of today – who will gradually take over the leadership of our institutions in the decades to come – have the knowledge, will, and capacity to become the change agents that are needed. Right now, many of them are struggling with student debt and making ends meet. If they are to effectively lead this effort to avert climate catastrophe, however, they will need to tap into the funding, professional experience, and networks that the boomers, Generation Xers, and all of us who have lived during the past seventy years now have to offer.

More than any other generation in history, the boomers have thrived during these past seventy years as the result of a growth-at-all-costs economic system and a personal value system that has assumed that "more is better." I believe we now owe something to the generations that follow us. They now have to deal with the unintended consequences of all that CO_2 spewed into the atmosphere from fossil fuels. Burning so much carbon has put their lives and the lives of their children in jeopardy. We now have the opportunity to give back – as well as to add greater meaning to our lives during retirement. More than ever before, now is the time for intergenerational collaboration and action.

What this book offers – and my hopes for its impact

In Our Hands is meant to provide something I feel is missing now: a short, easy-to-understand primer on global warming and a call-to-action handbook that provides specific actions everyone can take as well as educational materials for further study.

In order to paint a clear picture of the choices we face, Chapter 2, the *Probable Future,* describes how the world might move toward collapse over the next three and a half decades until 2050 if we deny global warming and succumb to the politics of fear. The *Probable Future* is admittedly a worst-case scenario. But even if it takes a few more decades to develop, this scenario is not unrealistic or overstated. This doomsday vision is not meant to scare or depress you, but rather to motivate you to get involved and take action.

In Chapter 3, a vision of the *Possible Future* describes how we can unite around values that will support the actions I believe we need to embrace if we are to avert the worst consequences of global warming.

Chapter 4 describes the *who* and *how* of the possible future – who stepped forward over the years to 2050 and what they did. Represented are generational groups such as millennials and boomers as well as many professional groups, including businesspeople, educators, and officials at all levels of government.

Chapter 5 directly addresses the question most people are asking: "What can I do?" The recommendations in this chapter highlight the many actions we all can take in our day-to-day personal lives, in our life as a member of our local community, and in our business and professional lives. At the end of Chapter 5, you will have the opportunity to write down what you are actually going to do and when you are going to start.

It is likely that elements of both the *probable* and *possible* future scenarios will be playing out in the coming decades: it is up to us to determine what the balance of disasters and successes will be.

Finally, the *Resources for Learning and Action* section is filled with suggested books, films, and organizations that pertain to many of the topics in this book, from the science of global warming and the shift to renewable forms of energy, to research on the evolution of values and consciousness.

If you become a part of the growing climate movement, you will experience the meaning of your life in a new way. It is no small thing to help avoid the probable collapse of humanity. Even more satisfying will be the new appreciation you will gain for your fellow human beings and for the natural world. You will be involved in a great quest, one for which your children will thank you. If you do nothing, what will you say to your children and grandchildren?

My hope is that if you are inspired by this book that you will seek to get it out into the world by sharing or giving copies to friends and family. Imagine what could happen if we all rose to the occasion and took action!

CHAPTER 1

Two Futures

There is an ancient curse: *May you be born in interesting times*

I fear that millennials, born between 1980 and 2000 or thereabouts, and the rest of us, are living this curse. There are any number of interesting upheavals underway in our society today, but none threatens the future of millennials, and the future of their children and grandchildren, more than global warming and climate change. The warming of the planet is happening more rapidly than our awareness of its potentially catastrophic consequences. My generation has wasted precious decades arguing about its existence, and the millennial generation (as well as others) seems generally numb to the existential nature of the threat. The word "existential" is not philosophical: it refers to the potential of climate change to wipe out much, or possibly all, of human existence.

My past and your future

Millennials are living in a world that no longer resembles the one I have lived and worked in, and thus the need for fresh approaches to the new challenges we are all now facing. I was born in the United States just before the Second World War unleashed horrendous destruction and sacrifice on a world that had already gone through a First World War and a Great Depression. After the war, everyone understandably wanted rapid economic growth defined in terms of the cars, refrigerators, housing, and other goods that would make their lives "better." And – much like today – governments, responding to near-term political pressures, wanted to do all they could to stimulate economic growth as a way to generate jobs and social stability. Consumerism, driven by advertising, thrived. At the time, nature's resources appeared to be inexhaustible. No one was concerned about the possibility of our economic activities warming up the atmosphere and threatening our very survival.

During this post-war period I was a U.S. diplomat seeking to extend America's values and power in an unstable world. For years I worked as a consultant for foreign governments, including Korea, Taiwan, Bolivia, and Saudi Arabia, as well as my own country, helping them develop industries that would support their economic growth. For years I worked for major corporations, such as Citibank and Toyota, helping them expand their global reach. For a time, I taught students at a U.S. business school about economic development, trade, and investment. All along, I felt I was a being a productive global citizen.

Until my "aha" moment two years ago in Ethiopia, I had embraced and promoted that economic system into which I was born. However, I had also witnessed how the fossil-fuel-powered burst of innovation and economic activity since World War II had added five billion more people to our small planet and created dangerous unintended consequences. As a result, I am convinced that market capitalism must now be harnessed in ways that avert further global warming and support healthy ecosystems as well as the well-being of the people in every country. In Chapter 4, I suggest several fundamental changes to how economic growth might now best be measured and market capitalism might best be harnessed to support the challenging work before us.

Whether you are 15 or 35, have you thought about what your life will be like 30-plus years from now? What will it be like for your children and grandchildren? In 2050 you may be in midlife or seeking security for retirement. Your kids or grandkids may be starting college or careers and families, longing for success and stability. I mean, really, think about this for a moment. Thirty or so years from now, what do you see?

We generally assume life is going to get better and better. Yes, we'll be driving electric cars – or they will be driving us. Yes, new technologies and artificial intelligence will transform every aspect of our lives. And, if we continue to burn the fossil fuels and engage in the many other human activities that carbonize our atmosphere, you probably assume the Earth's climate will be getting warmer and warmer.

The following chart depicts how CO_2 and other gasses are trapped within the sixty-two mile layer of atmosphere around the Earth and are the primary cause of global warming and climate change.

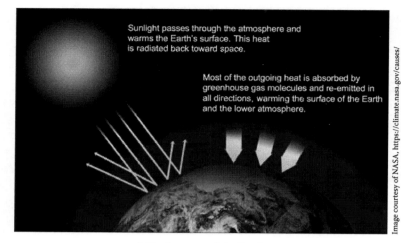

Sunlight passes through the atmosphere and warms the Earth's surface. This heat is radiated back toward space.

Most of the outgoing heat is absorbed by greenhouse gas molecules and re-emitted in all directions, warming the surface of the Earth and the lower atmosphere.

Image courtesy of NASA, https://climate.nasa.gov/causes/

A blanket around the Earth.

The graph on the following page notes how carbon dioxide emissions have risen to dangerous levels since the end of the Second World War.

The urgency of action

If people think about global warming at all, they envision a threat looming far into the future — a problem we can eventually get around to solving or mitigating. Governments are working on it, right? And, if governments don't get their acts together, new technological advances will take care of it. We can go about our business and not worry. Somehow it will all work out.

The reality is that catastrophic climate change is already upon us — and we are facing tipping points beyond which stopping disaster will be impossible. If you have doubts about the rapid warming of the planet and its consequences, I invite you to read the recent reports of the United Nations Intergovernmental Panel on Climate Change (IPCC) and

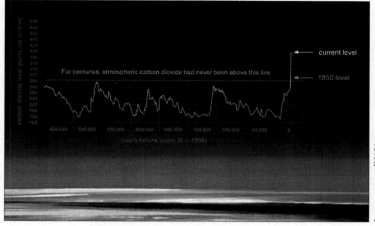

The relentless rise of carbon dioxide.

the American Association for the Advancement of Science (AAAS) and explore the many other materials you will find in the *Resources for Learning and Action* section at the end of the book.

A recent NASA summary statement reads: "Multiple studies published in peer-reviewed scientific journals show that 97 percent or more of actively publishing climate scientists agree: Climate-warming trends over the past century are extremely likely due to human activities" (https://climate. nasa.gov/scientific-consensus). Politicians may try and dismiss the science because of the financial power of the fossil-fuels industry, but I have much more confidence in science and scientists than politicians who are too often more focused on politics than policy. I also suggest we focus on the risk. If the catastrophic future described in the next chapter has only a 10 percent chance of happening, would you take that gamble? These 97 percent of all climate scientists believe the chances are more than 50 percent. How could we possibly take that risk and continue to debate with each other? I firmly believe that the new industries, new jobs, and economic growth generated by the shift to renewable forms of energy would far surpass that of trying to hang on to a fossil-fuel-driven future.

Not enough governments throughout the world are taking actions commensurate with the problem. This is a concern of mine for the United States in particular, where many political leaders either deny the problem exists or are not inclined to do anything about it. Despite the fact that over 50 percent of Americans say that they are either "alarmed" or "concerned" about global warming, the United States went through twelve hours of nationally-televised presidential debates in 2016 and climate changed was barely discussed. Many in political positions have made it clear that they will do all they can to gut the Environmental Protection Agency's rules on coal mining and relax higher gas-mileage requirements on new autos. It has made it clear that it will support gas fracking and an economic growth agenda driven by fossil fuels rather than renewables, such as solar and wind. In addition, many politicians in the U.S. seek to cut back funding in support of climate science and make it impossible for the United States to comply with the commitments made by President Barack Obama on behalf of the United States at the 2015 United Nations Climate Change Conference in Paris. As a result, the United States is passing on to China the global leadership efforts in stopping climate change and in developing renewables. And finally, insufficient attention is being paid by the media to the fact that we are at the edge of an existential precipice. We are, all of us, sitting like frogs in a pot of water that is coming to a boil so slowly we may not try to jump out until it is too late.

If we do not change our direction, we are likely to end up where we are headed.
—Ancient Chinese proverb

As I see it, we are making decisions every day at the individual, local, national, and international levels that will lead us to one of the following futures.

The Probable Future: In this future – made probable by the denial of human-generated global warming, the politics of fear, and inaction – we head down a path toward ecological disaster as well as economic and societal collapse.

I share this grim picture of our world in 2050 not to terrify and depress you but to give you a realistic view of what awaits us if we do not take actions, individually and collectively, commensurate with the threat. What many people have not yet understood is the reality of "tipping points" – occurrences that cause impacts from which we cannot recover, no matter what we do. In 2015, the Environmental Defense Fund identified six climate tipping points or points of no return: the melting of Arctic sea ice; the loss of Greenland's ice; the disintegration of the Antarctic ice sheet; the warming of the oceans; the drying out of the Amazon rainforest; and the loss of half of our boreal forests due to drought and fire. All of these points of no return are now advancing at an alarming rate.

The Possible Future: In this future – made possible by a transformation of values, collective positive action, and new technologies – we not only survive but benefit from the trauma with which all of humanity had to come to grips.

While global warming is a crisis that all 7.5 billion Earth inhabitants and governments in two hundred countries must address, the likelihood of the positive *Possible Future* actually happening depends greatly on what happens in the United States. Unless the U.S. becomes a leader of a revolutionary shift in human behavior and actions, I do not believe humanity will have much of a chance.

No one can predict the future with great accuracy, but we can identify the forces at work that will shape it. And we can track the footprints we take along the way down one path or another.

With this in mind, I would like to show you what these two futures look like and then ask you to consider these questions: Toward which future will you work? What specifically do you plan to do? When do you plan to start?

We will be known forever by the tracks we leave.
—Dakota proverb

CHAPTER 2

The Probable Future

How Our Actions Led to Our Collapse

In this future, humanity heads down a path toward ecological disaster as well as economic and societal collapse. The major driving forces that cause this tragedy are the denial of human-generated global warming; those in political power inciting fear of "the other," preventing cooperation among nations; economic policies that focus on fossil-fuel-driven growth; exploitation of the world's natural resources, and a consumer focus on the notion that "more is better."

Looking back from the year 2050

The 33 years between 2017 and 2050 have been the most turbulent and destructive in human history. Our missteps in some areas and inaction in others have led to our near-collapse today defined by the following disasters.

The world's temperature is more than 5.4° Fahrenheit (3° Celsius) above preindustrial levels. While some parts of the world were switching from carbon fuels to renewable energy in 2017, most of the world, including the United States, continued to burn fossil fuels until the mid 2020s when it was too late to avert the tipping point. By the mid 2020s the melting of the Arctic tundra had released huge quantities of the other greenhouse gas, methane, a chemical that had even more impact on global warming than CO_2. With much less ice in the Arctic each year, the sunlight that was previously reflected away from Earth was absorbed by the darker ocean. This sealed our fate. As had been predicted at the turn of the century, the planet had reached the temperature that would result in continued, irreversible global warming. The increased heat had made many places on the planet no longer inhabitable by humans and other species and had generated massive changes in weather patterns – with catastrophic consequences.

Air pollution levels have become deadly. Seventy percent of the world's 8.5 billion people now inhabiting our planet in 2050 live in urban areas where industry is located and most of the jobs are. But pollution levels are so high that most everyone's health is deteriorating and most are wearing masks. As a result, millions have been forced to move to rural areas where the air is breathable, despite the lack of jobs.

The average level of the oceans is now three feet above where it was in 2017. Melting ice from the Arctic, Greenland, and Antarctica has raised the sea level such that substantial parts of many coastal cities around the world are now underwater, and those that are still above sea level face more frequent hurricanes and typhoons with ever more substantial storm surges measured in the tens of feet. Miami, New Orleans, Boston, New York, Amsterdam, Shanghai, and Venice are now among those coastal cities most severely impacted. Most experts expect the sea to rise another three feet over the next thirty years.

Drought has destroyed agriculture and livelihoods throughout the world. While some areas of the planet became temporarily more productive, millions of people in Middle Eastern, African, and South Asian countries faced steady increases in the desertification of their farmlands, forcing them to become climate refugees with no safe places to go. The 129° Fahrenheit (54° C) temperature in Kuwait and Iran on July 22, 2016, soon became commonplace around the world. Millions died due to conflict or starvation. Fires consumed huge forests as well as towns and even cities in the western U.S. and Australia. Much of the Amazon rainforest, "the lungs of the planet," has dried up, contributing to continually changing global weather patterns.

Fresh water has become scarce in many parts of the world. The snowpack and glaciers of the Himalayan mountains that fed the great rivers of China, India, and Southeast Asia now provide a fraction of the fresh water they did in 2018 for inhabitants downstream. As a result, millions have died and millions more have become climate refugees. The glaciers of the South American Andes, where melting snow had for millennia supplied 80 percent of the fresh water to downstream populations, have also been reduced to a fraction of their former size with equally devastating consequences. As the American Southwest has also grown drier, conflicts over fresh water have become frequent and intense as farmers, urban communities, developers, and natural gas frackers battle for access over ever-diminishing quantities of ground and underground water.

The world's oceans have become acidified and have lost 90 percent of their marine life. As the planet's oceans — making up 71 percent of the Earth's surface and 97 percent of the Earth's water — grew warmer and warmer, they could no longer absorb CO_2, and their acidity rose dramatically.

Coral reefs around the world are dead and no longer provide habitat for spawning fish and sea creatures. Our oceans – overfished and a dumping ground for plastics and toxins – are turning into acidic "dead zones."

More than 80 percent of all species of animals alive in the year 2000 are now extinct. Climate change, overpopulation, and pollution have destroyed habitats, and poachers have killed off our most iconic mammals in the wild. Lions, tigers, elephants, bears, dolphins, and big apes exist mainly in a few zoos and "wildlife zones."

Climate refugees have no place to go and have died by the hundreds of millions. By 2020 the United States, Western Europe, and other developed countries had closed their borders to a mere trickle of refugees despite the reality that the number of climate refugees continued to grow rapidly. Now in 2050, climate refugees from the Middle East, Africa, South Asia, and South America, facing even more starvation and violence in their own countries, roam the planet destitute and hopeless.

Somali climate refugees at a refugee camp in eastern Kenya.

Image courtesy of Jerome Delay/Associated Press

International trade and investment are now a fraction of their previous size. Starting in 2018, when the U.S. erected tariff and non-tariff barriers to international trade, investment, and technology flows, other countries responded in kind in an effort to stimulate their own economies and protect jobs. When the U.S. rejected the Trans-Pacific Partnership (TPP) agreement and stepped back from its commitments under the Paris climate agreement, China took the leadership in both areas. Ultimately, all this disruption proved disastrous for individual countries and the global economy. The prices of commodities and consumer goods shot up in the U.S. and other advanced economies, which had been importing from lower-cost countries. The economies of the lower-cost countries declined as a result of their loss of export markets.

The U.S. economy is severely stressed, and the global economy is on the verge of collapse. Seriously stressed by ever-more-costly climate disasters, sea-level rise in major urban areas, resource shortages, agricultural failures, reduced international trade, and social unrest, the U.S. economy has become more and more unstable. Interest payments due on the massive amounts of debt that had long sustained economic growth could not be maintained in the U.S. and globally.

Vast economic inequality threatens the very fabric of life. The divide between the very few who can afford to live in fortress-like "gated communities" and the billions who now live in poverty has taken its toll throughout the world. The "haves" seek protection with private armies. Some cities are armed camps, and terrorist actions are everyday occurrences.

The "politics of fear" carry the day as demagogues and authoritarian power structures have gained power everywhere. Economic and social calamity have caused people to react out of fear and scarcity. Even in previously strong democracies, ruthless politicians have preyed on people's mistrust of "the other" for political gain. The global political environment has become ever more debilitated by short-term decision-making and vested interests holding on for dear life.

In sum, this is life in 2050. The warnings raised in the 2015 book *Overdevelopment, Overpopulation, Overshoot* have all come to pass, and humanity was not up to the task of addressing global warming before it was too late. Our failure in the United States traces back to two decisions made by political leaders in 2017. The first was to deny that global warming was a serious threat. The second was to engage in the politics of fear – instilling fear of "the other" rather than seeking ways to collaborate with the other.

Back to 2017... and hope for the future

This view of the future may seem far-fetched to many of you, but I feel comfortable in arguing that it is not. It is indeed a worst-case scenario; it assumes there is little change in our burning of fossil fuels. But even if we do manage to reduce carbon emissions, we will still be facing radical changes in the planet's climate. Only if the industrial world makes immediate, dramatic reductions in atmospheric CO_2 emissions can we avoid the outcomes pictured here. Unless we take action now, we will only postpone the inevitable – and these catastrophic changes will nevertheless create a horrifying world for our children and grandchildren.

By now you may be asking if it is even possible to avoid climate catastrophe? It is, but the window of time for action is getting smaller and smaller. Take a look at this graph from the IPCC that projects the likely warming of the planet if we do not take action, and if we do.

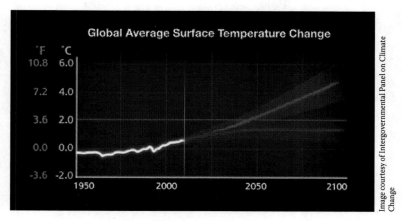

Projected global average surface temperature change depending on whether we take action or not.

The temperatures on the left side of the graph are in Fahrenheit and Celsius. A two-degree Celsius temperature rise is 3.6 degrees Fahrenheit. The upper swath on the top right projects the range of possible temperature if we continue the "business-as-usual" pace of carbon emissions we are currently emitting. The lower swath projects a leveling off of carbon pollution if we take the necessary actions.

As you can see, there is a pathway to prevent the consequences of the *Probable Future*. If the world slows the burning of carbon fuels sufficiently, we can avoid the worst consequences of global warming. There will still be very dangerous and uncomfortable changes in the global climate, but some mitigations will be possible. And who knows what future technologies (including massive reforestation) may be developed to actually remove existing carbon from the atmosphere in large amounts.

As we tackle the issue, we should keep in mind this famous quote:

> *No problem can be solved by the same level of consciousness that created it.*
> —Albert Einstein

Both Einstein and our current situation are calling us to embrace a new set of values – a new worldview that will drive a new series of actions that can transform our *Probable Future* into something more like the *Possible Future* described in the next chapter.

CHAPTER 3

The Possible Future

How We Saved Ourselves – A rEvolution of Values and Actions

In this future, made possible by a transformation of values, collective positive action, and new technologies, we not only survive but learn from our mistakes. The major driving forces that support this transformation are the realization and acceptance of climate change; working together to address global warming; redefining growth in ways that support both people and the planet; embracing sustainable sources of energy; and a recognition that "less is more."

Looking back from the year 2050

It took a near-death experience to wake us up, but we are now the better for it. The world is not "out of the woods," as they say, but the planet's temperature has not risen above the threshold scientists warned would make our planet uninhabitable. In many respects, the existential threats we faced head on, barely before it was too late, have brought us to this point of relative global stability.

The realities of global warming galvanized the human community into greater global unity, and its worst impacts have been averted. Humanity has made its peace with the natural world and is in the process of creating a sustainable and resilient ecological civilization.

The process of transformation was not without great suffering and pain. It was indeed a revolution – but it can be seen more accurately as a "rEvolution": an evolutionary step forward for humanity that resulted from both new actions and new values. Here's how it happened.

We woke up to the threat of global warming and stopped burning carbon

The most important step was that we woke up to the existential nature of the threat of continuing to burn fossil fuels, which include carbon dioxide, methane, and fluorinated gases. Between 2017 and 2020, numerous climate calamities occurred worldwide: The storms got more destructive, the droughts more severe, water levels began to rise in major cities throughout the world, and hundreds of thousands of migrants died fleeing parched regions of the Middle East and Africa.

Offshore wind turbines.

Finally, climate deniers no longer held sway, and governments and the media finally realized that we were facing a threat not unlike a global nuclear war – that continued warming could mean the end of the human species. Just as in the 1980s with the Soviet Union, and again in 2017 with North Korea, when the world stood on the brink of nuclear war, the United States in the 2020s finally helped lead a worldwide effort to curtail carbon pollution. Almost every other country in the world followed suit, enacting policies to reduce greenhouse gas emissions through the use of renewables, efficiency, and resource productivity. One of the most significant actions governments took was to reduce the direct and indirect subsidies the worldwide fossil-fuel industry had been receiving, which the International Monetary Fund estimated was $5.3 trillion in 2015, or $10 million a minute. The 2017 book *Drawdown: The Most Comprehensive Plan Ever Proposed to Reverse Global Warming* proved to be an invaluable resource for the 197 nations seeking to meet their commitments to the 2016 United Nations Framework Convention on Climate Change. *Drawdown* highlighted proven solutions not only in energy production and use but in food production, transportation, building materials, and many other areas.

Drawdown, as explained in the book, is defined as that point in time when the concentration of greenhouse gases in the atmosphere peak and begin to decline on a year-to-year basis. The book's goal was to identify, measure, and model the 100 most impactful, substantive solutions to global warming that either reduce emissions or remove greenhouse gases from the atmosphere by 2050. The 10 top solutions that the Drawdown initiative determined would have the greatest impact on reducing the greenhouse gases that are contributing to global warming are: refrigerant management (banning HFCs); offshore wind turbines; reducing food waste; shifting to plant-rich diets;

tropical forest restoration; educating girls; family planning; solar farms; silvopasture (the integration of trees and pasture into a single system for raising livestock); and the installation of rooftop solar panels.

Once awake, it was not hard to see that the age of burning fossil fuels was over. By 2017, the cost of alternative energy sources such as solar and wind had become lower than the price of oil, coal, and gas. For a few years, the political influence of the fossil-fuel industry tried to stop this switch to renewables, but the economics won out. There was no denying that the economic carbon bubble had burst. The fact that renewable energy projects also generated significant employment opportunities made it easier for politicians in the U.S. and elsewhere to support the shift to a "green economy."

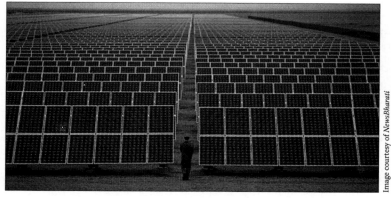

Image courtesy of NewsBharati

A solar farm.

A major factor in the transition to renewables was a carbon tax enacted in 2020. By then it had become clear that the fossil fuel industry created many costs to society that they did not pay for – specifically the pollution of the air, land, and water and the many other costly consequences of climate change. Recognizing that when you want an industry to pay for its impacts you tax it, a group called the Citizens' Climate Lobby lobbied the U.S. Congress to put a tax on the burning of carbon (much like the tax on cigarettes). The legislation they were able to get passed was actually a "carbon-fee-and-dividend system," a revenue-neutral carbon tax with 100 percent of the net revenue returned directly to households. People initially used their dividend to offset the cost of fuels, but it also enabled them to invest in electric cars, solar panels, new efficient appliances, and retrofitting their homes for energy savings. The new system had an immediate impact on greenhouse gas emissions, and was quickly adopted in a number of other countries around the world. In 2025, when China no longer allowed new gasoline or diesel-powered automobiles to be sold in the country, many other countries began to follow their lead.

As nations tackled reducing the source of carbon emissions, they also began to find ways to sequester carbon dioxide in the Earth with innovative reforestation, farming, and grazing practices. Most important was the preservation of the Amazon rainforest, where deforestation had been shrinking and drying out the forest for decades. Millions of acres were put into biological preserves and protected from farming and extractive industries. Massive tree-planting campaigns were carried out throughout the world. The revolution in agriculture sequestered carbon in the soil, and changes in the raising of livestock brought practices that mitigated the release of methane gas.

We rejected the politics of fear

Most crucial to the shift in values that changed the course of humanity was that we rejected the politics of fear. The process was a long and painful one, however, beginning in 2016 with "Brexit," the British vote to exit the European

Union, and a change in the political climate in the United States. It continued as other nations reverted to tyranny in addition to increased nationalism and racism. There were some dark years, as fear-based politics and short-term thinking seemed to be leading to collective disaster for everyone on the planet.

In confronting the impacts of these dangerous and reactionary policies, people throughout the world began to wake up. They began to realize that we had the choice to be driven by fear or by hope, possibility, and collective action. In response to fear-based politics, a non-violent revolution took place as more and more citizens realized they had to mobilize to protect their rights and prevent governments from becoming authoritarian, repressive, and regressive. The "battle" began to be waged at every level of society, as people of all ages and income levels joined forces to stop racist and reactionary programs and turn the tide toward the new, sustainable future they envisioned. These activists generated a spark that spread around the world, proving the truth of the famous quote:

Never doubt that a small group of thoughtful, committed citizens can change the world. Indeed, it is the only thing that ever has.

—Margaret Mead

We embraced higher human values

While there were many technological developments that enabled us to pull back from the brink, what was most crucial to our success was a shift in human values. We realized that we had entered the Anthropocene Age, a new period of the Earth's history when the impacts of humanity

Women's March on New York City, January 21, 2017.

on the planet became so significant as to fundamentally change our planet's ecosystems. As this new reality came into focus, we humans realized that we now had to steward the natural world rather than just take from it. That required a new level of cooperation and a greater sense of shared goals than ever before.

This new consciousness has also enabled us to recognize the emptiness of a life based solely on materialism, consumerism, and competition. People now find more meaning and fulfillment through connection with each other and the natural world. A new understanding of the oneness of all humanity has led to a greater commitment to social justice. We are gradually moving toward a world in which oppression is no longer tolerated and every human being has the freedom and opportunity to live the life they want.

In essence, we saw that many of the assumptions that had shaped our past were actually dangerous to our future. Our old values were not necessarily bad, but they had caused us to take actions that would ultimately lead to our collective demise if we did not change our ways.

Those who argued that it was impossible for Americans to shift values and actions quickly were reminded of how dramatically the government and people of the United States shifted to a war mentality after the Japanese bombed Pearl Harbor in 1941. Before the attack, over 95 percent of the American people had been against the U.S. becoming involved in the war. Yet, when the threat became real to them, they were willing to sacrifice and commit themselves fully to the war effort.

We redefined wealth, progress, and "the good life"

At the end of the Second World War, the nations of the world adopted Gross National Product (GNP) and its sister economic measure, Gross Domestic Product (GDP), as measures of each nation's economic health and growth, primarily by adding up the value of all the goods produced. Given the conditions at the time, that made good sense and, for the next seven decades, GNP became the accepted measure of each nation's perceived economic success or failure, regardless of the types of activities that were counted.

Over time, however, it became clear that GNP and GDP did not take into account the negative impacts of many of these "growth" activities on the world's ecological systems as well as on the lives of many people. For example, the clean-up costs of the BP oil spill in the Gulf of Mexico in 2010 were counted as positive economic growth even though the spill had done great harm to the marine life of the entire Gulf ecosystem. Nor did GNP or GDP take into account those activities that were positively impacting people, such as improvements to health, education, freedom, and happiness.

As a result of such realizations, economists and governments started focusing with increased urgency on new ways to redefine prosperity and national well-being. We came to realize the importance of distinguishing between good growth and bad growth.

As the global population climbed toward eight billion people – most wanting to share in the developed world's consumerism – it became clear that the demands on the Earth's resources were outstripping supply and destroying crucial ecosystems. Since the 1980s, the planet had been in "overshoot." By 2020, humans were using 60 percent more renewable resources than nature could provide. People woke up to the fact that this kind of economic growth and consumption was not even close to sustainable and that humanity could not survive the destruction of its natural support systems.[1]

After more than seventy years of living with the assumption that "more is better," many people also questioned whether consumption-oriented materialism was adding all that much to their lives. For decades, studies around the world had shown that beyond a certain level of income there was no correlation between money and happiness. Sometimes, well-off people were less happy, and less well-off people were happier.

---

[1] In 1985, Tarzie Vittachi, the Indian associate editor of *The WorldPaper*, of which the author of this book was the publisher, told the author this story: "One day in 1947 I had the opportunity to ask Mahatma Gandhi how he was going to satisfy the needs of the Indian people. Gandhi replied, 'I am not, my son. I am going to teach them to reduce their wants and satisfy their needs.'"

Others questioned whether the pressures of the modern world had caused their lives to be out of balance. Out of balance with each other. Out of balance with the natural world. They explored the wisdom embedded in the Hindu phrase Tri Hita Karana, used by the people of Bali, which means "to live one's life in balance with the natural world and with each other in community." They questioned whether living at "machine speed" rather than "human speed" was adding meaning to their lives.

As a result of such realizations, a great many people sought to move beyond their fixation on money, possessions, and speed to see other paths to happiness, such as being in nature, engaging with family and community, and being in service to others. But that did not mean that people had to adjust to a lower standard of living. Indeed, the quality of life improved for many more people as technological developments made it possible to continually "do more with less." In the 20th century, the great inventor, thinker, and futurist Buckminster Fuller had proposed that accelerating efficiency would enable greater and better outputs to be achieved with less inputs. At the beginning of the 21st century, a new understanding of nature's design processes brought dramatic changes to manufacturing. The biomimicry, "cradle-to-cradle," and "regenerative-design" movements created products that were ecological and recyclable and systems that were safer, more efficient, and essentially waste-free. A new, better consumerism flourished that did not harm people or the planet.

We began addressing the population challenge

The fact that the world's population had increased by six billion people in the 70 years since the Second World War was a major wake-up call, calling into question whether such growth was sustainable. It was not just a matter of the absolute number of people, but also the expectations all 7.5 billion of us held for "more" – for the material possessions that the developed world took for granted. Continuing to provide more and more goods for more and more people would do even greater damage to the natural world and the atmosphere. The millions of migrants fleeing starvation and violence in their native lands became daily examples of this seemingly intractable problem that would only get much worse if global warming and overpopulation continued.

The subject of curtailing population growth in areas that could not support it finally burst into public discourse in the 2020s. It was increasingly recognized that the world had more people than the Earth could sustain, and that millions would die. This realization demanded a moral response to support those who already inhabited the Earth while seeking ways to slow down the rate of population growth. Since populations in most advanced countries were already in decline, the key question was how best to support the health and stability of those in poorer countries where having more children was seen by parents as a form of security. When women knew that more of their children would survive, they had fewer children, and greater access to birth control made this possible. It was also well known that investing in the education of women in the poorer countries would lead to lower birth rates, improved family health, and better standards of living. By 2025, the education of girls became the accepted norm in most countries of the world to indirectly yet effectively address the need to slow down the growth of the world's population.

We moved toward connection and cooperation

Perhaps the overarching transformation in values that took place was our recognition that we all needed to work with each other to survive and prosper. "One world, one people" seemed a naïve dream at the beginning of the 21st century, but as climate change knew no boundaries, nations had to come together to prevent collective catastrophe. As countries began to collaborate to reduce CO_2 emissions and implement policies that might mitigate the impacts of global warming, a new era of international cooperation was launched. It began with a new understanding of humans as borne out of the Earth, connected to everything, and responsible for the future evolution – or devolution – of the planet and our species.

This new consciousness reflected the ancient wisdom of many religious traditions as well as the profound knowledge of indigenous people who embodied this worldview in their teachings and way of life: the Earth is our Mother, and we are all one interdependent family. The profound wisdom of their message to the modern world, which we seemed to have forgotten in our headlong rush to modernity, was finally appreciated by more and more people. We realized that we had to work in community rather than solely for our own ends.

The shift from separation and competition to connection and cooperation was perhaps the most powerful factor in saving our planet, and it represented what might be called a New Story for humanity. The Old Story saw humans as separate – from the natural world and from each other – and competing for scarce resources in a world of winners and losers. The New Story of connection and collaboration did not mean there would be no more competition.

Capitalism was transformed such that cooperation was encouraged where it would protect and restore the ecological health of the planet, and competition was aimed at what could create the greatest good for the most people.

As was the case when the Soviet Union suddenly collapsed in the late 1980s, it had been hard to see the small changes in attitudes and behaviors taking place under the surface until one day we realized that we were into a new era of human evolution. It indeed was a "rEvolution" in our consciousness and actions and it spread from a small percentage of the world's population to most of us.

At the beginning of the 21st century, the Pachamama Alliance, an organization based in San Francisco, California, had articulated a vision for the future that began to spread worldwide:

To bring forth an environmentally sustainable, spiritually fulfilling, and socially just human presence on this planet.

This grand vision captures the essence of the world that we are heading toward in 2050. Although we are still far from fully achieving that vision, a new consciousness is guiding the human family toward a sustainable, just, and thriving world.

CHAPTER 4

The Possible Future
Who Made It Happen and What Did They Do?

Our success in steering the world away from global warming and toward a resilient and sustainable future was not the result of some formal plan developed by world leaders and handed down to governments, organizations, and individuals to act on. Far from it. It was more of an organic, grassroots process driven by millions of deeply concerned individuals, NGOs, businesses, and governmental organizations experimenting their way toward a positive future. Humankind finally realized that we shared a common goal — to do all we could as fast as we could to stop global warming. We realized we had to stop working at cross purposes and start working as a team, whether we were government officials, businesspeople, NGO activists, or people involved in any of the other professions that humans are engaged in.

Two age groups in particular stepped up to provide leadership.

A "rEvolution" led by millennials, supported by boomers and all of us

The path to the positive future we now enjoy in 2050 was driven in large measure by millennials. They were the children of baby boomers, whose numbers they surpassed in

2015. These young adults were the most educated generation ever, and they recognized the dangers a warming planet had on them and their children. Born into the greatest technological revolution in history, they were tech savvy and connected through the Internet and social media. In the United States, they were also the most diverse generation and the most tolerant of diversity.

Many millennials questioned the 20th century values into which they had been born. While a 2014 *Time* magazine article called them the "Me, Me, Me Generation," a report by the Pew Research Center described them as "confident, self-expressive, liberal, upbeat, and open to change." They were highly socially conscious and entrepreneurial, experimenting with new ways of living. They were suspicious of advertising that pushed unneeded products on gullible consumers. They believed that the future called for less focus on material wealth and more on the sharing of resources, such as houses and cars. As the U.S. government embraced more authoritarian and reactionary policies in 2017, they grew more politically active. And as the global climate situation grew more dire, they became more active in the movement to transition away from fossil fuels and toward a sustainable future built on renewables and doing better with less. In their passion for activism, they shared kinship with the baby boomer generation that came of age in the legendary Sixties of the 20th century.

The millennials' motivation was not wholly altruistic. While they were aware of the threat global warming posed for everyone, they also wanted to live comfortable lives – but on their own terms. They wanted to live lives of meaning as well as of relative comfort, albeit in ways different from those of their parents. That did not mean they were necessarily at odds with their parents, but rather that they could see that the "more-is-better, economic growth-at-all-costs" ways of being were not sustainable, and maybe not that meaningful. To be fair, many members of the baby boomer generation had long before moved beyond lives focused primarily on material wealth, although they were the exception.

In 2017, many millennials were struggling with the financial stresses of student debt, low pay, employment uncertainty, and a rapidly-changing job market, but many were moving into position to be the leaders of the professions. They knew they needed all the help they could get in terms of financial resources, connections, and experience, and that's what their parents and other boomers had. They reached out to the boomers to build intergenerational collaborations for the future of humanity – and the rest, as they say, is history.

Why the boomers?

In the United States, the boomers were the generation born during the post–World War II "baby boom" between 1946 and 1964. The baby boomers had many reasons to collaborate with the millennials. They realized that they had benefitted greatly from the extraordinary economic growth brought about by the fossil fuels that were now the source of the world's greatest problem. And for this reason, many boomers felt a moral responsibility to support the transition to a fossil-fuel-free world. Many had financial resources the millennials needed in order to succeed. Many had pertinent expertise and contacts. They also had the time to devote to activism and intergenerational collaboration, since many of them were now retired and most could expect decades of relatively good health. Many were ready for a powerful partnership. A challenge for some boomers was to step back and let the millennials lead, a reversal of roles that did not come naturally.

The collaboration between millennials and boomers began to make headway in 2018. One particularly successful initiative took place in San Francisco where a group of millennials, supported by members of the boomer generation, began working together to create a social network to galvanize the climate movement. By the spring of 2019, they lit a spark that launched what became known as the Global Spring Movement. Using Facebook, Twitter, and their new social network, they galvanized other millennials, and by the end of 2022, over 200 million people in 112 countries were on board, exploring solutions and taking actions in their lives and professions.

The movement grew rapidly in number and influence. While in 2017, only three percent of the world's population was estimated to be deeply committed to addressing global warming and climate change, that number grew to five percent in 2019 and to 10 percent by the end of 2020. That was enough to further influence and galvanize other elements of each country's population. This growth proved the accuracy of the well-researched theory of change that a small number of passionate, well-organized, and focused individuals can bring about fundamental change if they have 10 percent of the population so engaged. Resistance from vested interest groups and ill-informed individuals continued, but their numbers diminished over the ensuing decades. The millennial-boomer coalition was a major factor in this societal shift.

Individual action: the power of many "ones"

Prior to 2017, despite the fact that over 50 percent of Americans in polls expressed serious concern about global climate change, few discussed it or took significant steps in their daily or professional lives to address it. Starting in 2017, however, the political climate galvanized many individuals to become far more active. They began to explore how they were contributing to global warming and how they could be part of the solution. People realized that it was not only in their enlightened self-interest to change the ways they were living, but that it was also an exciting and meaningful challenge to take on in their personal and public lives. In a world filled with numerous challenges, all seeking to be top priority, addressing climate change became an organizing principle.

The simplest and most potent action they took was that they talked about climate change. Individuals began raising the issue of global warming and what to do about it with their friends, family members, and professional colleagues. Many started "global warming discussion circles" along the lines of church groups, book clubs, and movie groups that had for years explored other topics. In fact, many church and book groups shifted to global warming as a topic to explore. Gradually these groups invited others who did not share their views on climate change to join their discussion circles. One of the rules was that "discussion" did not mean having arguments, but rather the development of deep listening skills and a search for common ground and collaborative action for the common good.

Many individuals joined existing activist organizations such as 350.org, the Sierra Club, the Climate Reality Project, NextGen America, and Greenpeace to participate with others and physically show up – whether at a rally in their own town or in a major city such as New York or Washington, D.C.

Image courtesy of Oregon Just Transition Alliance

Oregon Just Transition Alliance march in Portland, Oregon, April 29, 2017.

As political leaders began attempting many unpopular policy changes with regard to fossil fuels, the environment, civil rights, immigration, and healthcare, public protests became more prevalent. People who previously were content to sign online petitions began to show up for peaceful demonstrations. A recent model of non-violent protest had been modeled by the Native American tribes and their supporters who attempted to stop the Dakota Access Pipeline in 2016. Though their victory in North Dakota was reversed by the new administration in 2017, the movement, led by indigenous activists, spread throughout the United States as people everywhere began to organize to stop pipelines, coal plants, and other pollution problems.

Individuals also made conscious choices to change their lifestyles – such as shifting away from fossil fuels and embracing renewable forms of energy in their homes, cars, and businesses. Aware that over 30 percent of our food supply is wasted – left in the fields, spoiled in transit, uneaten at the table – they changed their eating habits, cutting down on meat and consuming more locally-grown produce that reduced transportation carbon emissions. They bought less stuff and reused and recycled more, and they shared more – bicycles, cars, even houses. They chose carefully where they purchased their goods and services, who they worked for, which companies they invested in and divested from, what media they read or watched, and who they voted for. Businesses, investors, and politicians all took notice and in turn made changes that were responsive to these shifts in priorities and behaviors.

From ME to WE

It was clear, however, that actions taken as individuals – important as they were – were not going to have the impact needed to address a problem the scope of global warming. As climate activist Bill McKibben had put it, everyone needed to go from thinking about ME to thinking about WE – what we could and had to do collectively if we were going to turn the tide. Here is how a wide range of groups, businesses, organizations, and government entities took on the greatest challenge of all time – and succeeded.

Women's organizations

In 2009, the Dalai Lama said, "The world will be saved by the Western woman." The truth of that quote became clear in the 2020s and 2030s when women in the developed world attained full equality with men. A new wave of feminism began in response to President Donald Trump and the #MeToo movement as women organized, marched, ran for political office, and took their rightful places in the world of business. A woman president soon followed, and the feminine perspective began to flourish as men realized that it supported and nourished them as well. Western women did not stop there, however. They pushed for radical change for women throughout the world, ending the worst practices that injured and hampered women, especially in the Middle East and Africa, implementing supportive policies and cultural changes in societies throughout the world.

Nonprofit organizations

For decades, thousands of nonprofit organizations and thought leaders had sought to bring about changes in humanity's goals, values, and actions to achieve a more sustainable, resilient, and just world. As people became more aware of the climate crisis, these groups redoubled their efforts to bring about a global agenda of environmental and social programs that would not only address global warming but would also improve the lives of billions of people. Examples of their initiatives included: protection of biodiverse territories, organic farming, reforestation, soil restoration, permaculture, the recycling of plastics, support of indigenous cultures, the education of girls, maternal and child health, and family planning.

Organizations such as the Environmental Defense Fund, the Natural Resources Defense Council, Friends of the Earth, Greenpeace, 350.org, the Climate Reality Project, the Indigenous Environmental Network, Rainforest Action Network, Bioneers, and the Pachamama Alliance were just a few of the thousands of organizations in the United States and around the world doing extraordinary work, often far from the headlines but with powerful results. NGOs such as Encore.org, which focused on the many ways older Americans could add meaning in their retirement years, jumped at the opportunity to get the boomer generation to support the millennials in taking action on global warming.

Businesses

To many, the response from the business world to the challenge of climate change was surprising. Early in the 21st century, many major businesses had realized that it was in their self-interest to act, and they proved more agile than imagined. They could see the spiraling costs of operating in climate chaos. They knew that they would not only lose money, but also customers, talented employees, and investors if they did not authentically address the issues of reducing carbon emissions and the efficient use of resources. The millennials in particular were quick to call out corporations that sought to continue business as usual.

Corporations were able to see business opportunities when others bemoaned the challenges that change would bring. For example, wind turbine power generation expanded exponentially, as did solar farms and the manufacture and installation of rooftop solar panels. Corporations in the refrigerant business, which by international treaty were required to stop using HFCs, developed environmentally-safe alternatives. Cement manufacturing, which used massive amounts of electricity, shifted to materials that were more energy-efficient.

Antitrust laws were modified to enable competitors to collaborate in areas that would support a shift to a more resilient and sustainable economic and environmental future. At the same time, businesses were left to compete with each other for customers, employees, and investors, as had long been the case. Once the big corporations got behind the renewable economy, politicians started paying attention.

Entrepreneurs and intrapreneurs

Social entrepreneurs and small businesses also played important roles. Social entrepreneurs are individuals who seek to address social or environmental challenges by using business methods to make projects self-sustaining financially rather than relying primarily on philanthropic contributions. For example, they devised innovative methods of delivering small solar and wind-driven energy and fresh water throughout developing countries. They came up with significant innovations to deliver better quality health and educational programs.

Intrapreneurs are employees of organizations who seek to improve the processes or products of the companies for which they work. Instead of protesting against a company's practices from the sidewalk they actively tried to push for changes from within. They were not seen as whistle-blowers. They were increasingly seen as important change agents within their companies.

Solar panels on African huts developed by social entrepreneurs.

Investors

By 2017 many long-term investors were clear that the shift away from fossil fuels to renewable sources of energy was the way of the future – not only because of impending climate catastrophe, but also because the price of renewable energy was already an attractive alternative and would become ever more so. Investors had also noted that the more-aware oil companies had already started to think of themselves as "energy companies" in recognition of where the future was heading. Long-term investors were also aware that consumers were increasingly turning their backs on fossil fuels. Many institutional investors had already decreased their holdings in fossil fuel companies, just as they had in tobacco stocks and in companies invested in South Africa during Apartheid.

Increasingly, companies were graded by their Environmental, Social, and Governance (ESG) criteria, not just by their projected financial returns. ESG referred to how a company or corporation took into account environmental considerations, practiced social consciousness, and governed itself. Included were markers such as equal opportunity for advancement, the percentage of women in top positions, and the ratio of lowest paid to highest paid employee. By the mid 2020s, impact investing, in which individual investors sought to consider both financial return and social good to bring about social change, had also become the norm.

Architects and builders

The entire construction industry – architects, developers, builders, and building-products companies – experienced a renaissance as new energy-efficient design and resource-efficient materials made possible zero-carbon buildings.

Green buildings, using structures and processes that were environmentally responsible, became the norm. Homeowners sought to change their energy sources, roofs, windows, and lighting. Alternatives were developed for cement and other highly energy-intensive products. Plastics, a 20th century boon to most industries and consumers but a growing scourge to the planet, were addressed by advances in recycling and a shift in consumer attitudes. The era of one-time use of "disposable" plastic products came to an end as people who were accustomed to throwing things away grappled with the question, "Where is away?"

The legal profession

Throughout the societal transformation, lawyers and activist legal organizations, such as Earthjustice, the Gaia Foundation, the Earth Law Center, and the Global Alliance for the Rights of Nature, played a significant role in supporting policies and actions to slow and stop global warming. One of the most important and effective of these was Our Children's Trust, which represented young people in a series of lawsuits against the U.S. government and the fossil-fuel industry. The young plaintiffs asserted that the government's actions in causing climate change violated their generation's constitutional rights to life, liberty, and property; their future was threatened. In 2017, in a landmark case in Oregon, the federal court agreed that the constitutional rights and public trust rights of young people were being violated, which paved the way for the future breakthrough Supreme Court decision that CO_2 emissions must be regulated.

Scientists

Since the turn of the century, scientists had played an increasingly important role in bringing the realities of climate change into public discourse. The Union of Concerned Scientists, an alliance of more than 400,000 citizens and scientists, had pushed for government policies, corporate practices, and consumer choices that would slow global warming. The role of the scientific community became even more crucial during the time when science was under all-out attack, with governmental research being shut down, its funding cut, and scientists silenced. The pushback was enormous, as scientists – not usually at the forefront of resistance – organized and took bold actions to preserve their work and the truth about global warming. Their efforts paid off when the government changed in January of 2021, and the work of shifting to a fossil fuel-free economy began in earnest.

Mainstream media, documentaries, and social media

At the beginning of the 21st century, the mainstream media was slow to recognize and report on the climate crisis, and climate change leaders such as Al Gore, James Hansen, Leonardo DiCaprio, and Bill McKibben often seemed like voices crying out in the wilderness. But, as the calamities grew, there was no denying the newsworthy nature of the strange weather patterns and their tragic consequences. Media coverage increased and many in the entertainment industry began raising the issue in informational films.

Documentary series such as *Years of Living Dangerously* and films such as *Chasing Ice, Before the Flood, Merchants of Doubt, This Changes Everything, Climate Refugees* and *An Inconvenient Sequel* began to reach wider audiences. Documentary film producers, such as the Sacred Land Film Project, and the Goldman Environmental Prize, an international environmental awards program, gained greater recognition for highlighting both the challenges and success stories of people seeking to deal with the destruction and restoration of their environments.

Social media was a particularly powerful tool, enabling information and ideas to be exchanged around the world in seconds. A positive new vision emerged and began receiving widespread attention and support.

Educational institutions

Schools, colleges, universities, business schools, and trade schools gradually introduced new curricula for the new age we found ourselves living in. They offered courses on the scientific and social aspects of climate change and focused research on innovative ways to stop carbon pollution and mitigate the destruction brought about by climate change.

Organizations such as NatureBridge, an organization that for decades had been teaching environmental science and stewardship to young students in U.S. national parks, experienced a significant uptick in interest. And in the 2020s, college campuses became the centers of protests against global warming, not unlike the anti-Vietnam war and pro-civil rights student protests of the '60s and '70s. Ultimately, they forced most universities to divest themselves of ownership in the fossil fuel industries.

Religious and spiritual groups

A great boost to the movement for change was the ground-breaking encyclical released by Pope Francis in May of 2015. *On Care for Our Common Home* called for a new partnership between science and religion to combat human-driven climate change. The Pope called upon all Catholics to take the issue seriously and take action. Leaders of other faiths, including Islam, also joined the call, and interfaith religious groups, such as the U.K.-based Alliance of Religions and Conservation and the U.S.-based Interfaith Power & Light, continued their work to mobilize people of all religious faiths to join together to address the threat of global warming.

Organizations such as Earth Charter, the Institute of Noetic Sciences, the Shift Network, Wisdom 2.0, Commonweal, NewStories, and hundreds of others around the world that suggested new ways of imagining the future of humanity and consciousness became more and more mainstream. People increasingly delved into the work of pioneers in these fields, including the Dalai Lama, Desmond Tutu, Wendell Berry, Barbara Marx Hubbard, Jerry Jampolsky, Duane Elgin, Marianne Williamson, Charles Eisenstein, and many others.

Agriculture and food industries

Industrial agriculture and the food industry went through massive changes as people became aware that current animal production methods were right up there with transportation as a major contributor to greenhouse gases. According to soil and water specialists at the University of California Agricultural Extension, 5,214 gallons of water are required to produce a pound of beef while only 24 gallons are required to produce a pound of potatoes, 25 gallons to produce a pound of wheat, and 815 gallons to produce a pound of chicken. Assuming one takes a seven-minute shower each day, averaging 14 gallons of water each time, one can save more water by not eating a pound of beef than by not showering for a year. They also addressed the fact that up to 35 percent of food in high-income countries was being thrown out by consumers and that a third of food production in low income countries with poor transport infrastructure and refrigeration was being lost between rural farms and the urban consumer's plate.

Locally grown food production had a resurgence as a way to reduce fossil-fuel emissions resulting from long-distance trucking and long airline flights. Organic foods became the preferred choice by those who could afford it, not only for health reasons but because organic food production did not use petrochemical-based fertilizers. The Slow Food movement, which began in Italy in 1986, had chapters in more than 150 countries by 2017 and continued to expand rapidly.

An entirely different – but not new – approach to farming gathered momentum, as young people returned to the land to farm in both old and new ways. Using the principles of permaculture and other innovative approaches, these young farmers found financial success in small farms catering to new consumer tastes.

And, to the astonishment of many, in 2022 the U.S. Food and Drug Administration (FDA) issued new food guidelines recommending that individuals substantially reduce the amount of animal products they consumed. While this recommendation was in part due to health concerns, the announcement made it clear that beef and pork production – particularly CAFOs (confined animal feeding operations) – had for decades been adding significantly to global warming. They noted the large amounts of methane cattle excrement releases into the atmosphere. They highlighted the large amounts of fresh water cattle consume per pound of beef produced and the vast amounts of land required that could better be used to grow other foods using a fraction of these resources. As more "meat-like" new food products became available, it was easier for people to shift to a primarily plant-based diet.

Transportation

Major investments were made in the development of electric cars and trucks, high-speed rail systems, more fuel-efficient airplanes and ocean-going ships. At the local level, the public's increasing commitment to address the threats to their lives from increased CO_2 emissions caused them to embrace mass transit, walkable cities, ride sharing, the use of traditional and electric bikes, and mass transit systems.

States, cities, towns, and local communities

Because the national government in 2017 pushed back on climate change and progressive ideas, state governments became much more involved in promoting climate policies – as well as in protecting civil rights and progressive economic measures. California, the world's sixth largest economy, set the standard for state-mandated climate action. The state became a model for other states, which became more progressive as national regressive economic and political policies failed. Hawaii was also at the forefront of state action, meeting its target to achieve 100 percent renewables by 2025.

People took to heart the expression "Think Globally, Act Locally." Cities, towns, and communities were where many climate challenges, such as flooding from rising ocean waters and shortages of fresh water for agricultural or home use, had to be addressed. They started demanding clean-energy alternatives from local utilities. They pushed for plans that mitigated rising sea levels. They took the lead in developing local resilience and sustainability plans. These included new urban and suburban designs, energy-efficient buildings, improved public transportation, sophisticated recycling centers, community living, local farmers markets, public gardens, and living spaces that supported healthy lifestyles, to mention just a few. The work being done at the local level served as incubators and models for action at the state and national levels.

National governments

Ultimately, a former British Prime Minister's quote in December of 1941 proved prescient.

Americans can always be counted on to do the right thing – but only after they have exhausted all other possibilities.

—Winston Churchill

Since the turn of the 21st century, the United States military had been concerned about the dire social, political, and military impacts climate change would likely have on the world as a whole. They were also concerned about its impacts on their many military bases in coastal areas where sea level rise and storm surges were a direct threat to such installations. The military leaders realized that global climate change would result in fights among people over land and resources and that mass migrations would destabilize whole countries and regions. Combat commands integrated climate-related impacts into their planning. The Defense Department's response to climate change was way ahead of the rest of the U.S. government.

Until the 2020s, the U.S. government lagged behind not only its own military, but also many Western European governments. It took some years for politicians – ever-focused on the near-term wants of their constituents and the politics of the moment – to take the long view and recognize that disaster was looming. Significant action finally came when the U.S. Congress moved to reduce carbon emissions by putting a price on carbon through a carbon-fee-and-dividend system. In addition, they phased out all subsidies for oil, coal, and gas production and reallocated funds to support the rapid transition to renewable sources of energy.

In 2023, the U.S. Congress established the Global Stewards Service Corps, a mandatory two-year service program for all young adults. All young men and women who did not volunteer for the U.S. military were required to work for programs such as the Peace Corps, Mercy Corps, Habitat for Humanity, or Teach for America. The rationale for the Global Stewards Service Corps was both to get the work of stewarding the planet done, but also to refocus the lives of young Americans toward greater service to their communities. Congress also established the Green Corps, a program for at-risk young adults to be trained in installing solar panels.

Around the same time, GNP and GDP measures of each country's economic growth were overhauled to reflect the impacts of economic activities on the sustainability of the planet and the welfare of people.

A key challenge to all governments – not just the United States – was how to move to an economic model that would generate employment and social stability. In part, governments did this by encouraging investments in new industries, including renewable energy and in repairing and replacing existing aging infrastructures, employing workers whose jobs were lost in the transition away from oil, coal, and gas production.

The United Nations

The United Nations had, of course, been convening global climate summit talks since the 1990s. Finally, the climate summit in Paris in 2015 resulted in a 197-nation consensus about goals, timelines, and mechanisms to prevent a two-degree centigrade rise in global temperature – what had been determined to be the goal for stopping runaway climate change. In late 2016, they solidified these gains by getting the required 55 nations representing 55 percent of global emissions to formally commit to the agreement.

On November 17, 2017, the 2016 Kigali Amendment to the Montreal Protocol to phase out hydrofluorocarbons (HFCs) by 2028 entered into force. HFCs are the chemicals most widely used in air conditioners, refrigerators, and other cooling equipment throughout the world that scientists estimate are up to 9,000 times more potent than CO_2 to warm the atmosphere. Scientists had also estimated that this binding international agreement, if fully adhered to, would reduce global warming by one degree Fahrenheit by 2050, the most impactful of all actions that humans can take to address global warming.

Throughout the next decades, as nations around the world embraced the need for coordinated action on global warming, the United Nations was finally able to foster greater international cooperation toward that goal.

Now, in 2050: a more resilient, sustainable, and prosperous world

The transformation that took place between 2017 and 2050 came about not only in response to global warming but also as the flowering of a movement to create what author Charles Eisenstein called "the more beautiful world our hearts know is possible."

In 2017, many people had given up on a positive future for humanity. Popular culture was rife with books, movies, and television shows about a dystopian future in which the Earth was ravaged, totalitarianism or violent chaos reigned, the chasm between the very few very rich and the very many very poor was vast and deep, and only robots and artificial intelligence could save us.

Despite this discouraging cultural fantasy, there were those who envisioned a different future and planted seeds of hope and tended fields of innovation. First, they began by acting to stave off the worst policies of governments bent on dangerous short-term policies. They acted to safeguard wild places and plants and animals threatened with extinction. They fought battle after battle to finally win the war against pollution of the atmosphere. Even in the darkest days, when dystopia seemed about to swallow us, they did not give up.

Understanding the crisis we faced, knowing the work was on behalf of all of life, people gave their all to the healing of the environment and the human family. They looked ahead with the Native American vision of Seven Generations: that all decisions and actions must take into account the welfare and well-being of seven generations to come. So their plans were ambitious and their actions were bold. They knew the rEvolution they were making had to be a collective effort, so they came together as never before – in communities, online, and in the streets – standing, marching, and working for the more beautiful world they knew was possible.

Now, at mid-century, many of those heroes have passed on, but many remain to see us through difficult times and to continue to envision and create an environmentally sustainable, spiritually fulfilling, and socially just human presence on this planet.[2]

Today, more than ever before, life must be characterized by a sense of Universal responsibility, not only nation to nation and human to human, but also human to other forms of life.

— His Holiness the Dalai Lama

---

[2] And Kala, the Ethiopian you heard about in this book's Preface, is now sixty-two. He had to flee his country in 2016 in the face of imprisonment and very possibly torture and death because he had spoken out about the way global climate change coupled with his govenment's economic growth-at-all-costs policies were destroying the lives of many of the tribes in the South. In 2025, the Ethiopian government, as a result of international pressure and climate realities, granted amnesty to Kala and all those other Ethiopians who had protested and survived.

CHAPTER 5

What Specific Actions Are You Going to Take?

Become a climate activist

My definition of a climate activist is someone who is clear that the warming of the planet is a serious threat to the future of humanity and is committed to taking substantial, concrete actions to become part of the solution.

If you choose to be a climate activist, you can choose from a range of actions: Some could be public and all-consuming, such as running for local, state, or national office to address the issue legislatively. Or your actions could be private, such as working to lower your own carbon footprint. This chapter suggests the many ways you can take actions not only in your personal life, but in your life as a member of your community and in your professional life. Whatever you choose to do, the key is to make conscious choices and then follow through.

There are plenty of important tasks to go around, so you might as well choose activities that you are good at and will mean the most to you. Deal with the plastic issue? Organize a climate action group in your community? Try and shift your company's use of energy to renewables? Put your knowledge as an engineer to work to address how to protect your community from rising seas? The sections below provide many suggestions. You might also look again at the actions mentioned in Chapter 3 and Chapter 4.

Be aware that becoming a climate activist may well mean getting out of your comfort zone – stretching yourself to let go of old habits and embrace new actions. It may require that you make different financial choices based on new values – solar panels on the roof or an electric car may be more rewarding than flying off on a foreign vacation or buying an RV or boat. You may feel called to let go of eating beef – or any animal products – in favor of locally grown plant-based organic foods as a way to improve your own health as well as the environment. You may even risk arrest in a direct action aimed at changing climate policy. Whatever your "stretch zone" may be, it is likely to be challenging, stimulating, and rewarding!

Below is an overview of some of the actions you may consider in your personal life, your community, and your business or workplace. Included are some questions to ask yourself as you read, and a space below each section to jot down answers or comments as you go.

Educate yourself

If you feel you need further information and education about the issues, policies, and solutions related to global warming, look through the many *Resources for Learning and Action* that follow and dig in! There you will find many very informative documentary films, books, and organizations focused on most of the topics raised in this book. Also consider signing up to receive the newsletters of groups that are addressing topics of interest to you, such as the one produced by the Environmental Defense Fund.

What do I want to learn more about?

_____

_____

_____

_____

_____

Examine your values

Take a look at the values you have that may be driving actions that contribute to the problem. Are you seeking happiness primarily through buying and consuming products that advertisements tell you will make your life better, without considering their impact on people and the planet? What would a life of meaning look like for you that is not oriented around the consumption of resources? Being in nature? More time with friends and family? You may need some help with this process, as sometimes we can't see our own unexamined assumptions. Seek support from like-minded friends and family, or from a church or other group.

What values do I want to look at changing?

_____

_____

_____

_____

_____

Talk about global warming

Because the dangers of climate change are seemingly so far off, so frightening, and because so few people understand them, many people are still unaware or in denial. Thinking there is nothing they can do about it, they just don't want to think about it, and they avoid talking about it. This needs to change. Be brave and start conversations with your friends and family members. Ask how they think and feel about global warming and why. Then listen! You don't have to go off into lecturing or teaching, just engage and see what happens.

Who are the people I can talk with about global warming?

_____

_____

_____

_____

_____

Create a discussion and action group

Invite a group of friends or colleagues to start a discussion group, and set up regular meetings to discuss what you will commit to individually and collectively. You may choose to make your group *intergenerational,* bringing together young people who now need to lead the movement with elders who can support their efforts with funding, contacts, and experience. You will find some great information about how to create and run "engagement circles" in the *Resources* section. You might also go to www.InOurHands.Earth, buy a copy of this book, and take advantage of the offer to give up to five copies of the ebook to other members of your discussion group for free.

Who might I invite to a discussion/action group and how might I organize it?

_____

_____

_____

_____

_____

Speak up to power—and vote

Claim your rights as a citizen to influence public policy and legislation by regularly communicating with your elected officials. This is particularly important now that our democratic process is being threatened by climate deniers, the power of big money, by gerrymandering, and by voter suppression. Support legislators who reflect your views and challenge those who do not. Signing online petitions (sometimes called "clicktivism") can be useful, but it is not enough. Make phone calls, write real letters, show up at events to let your representatives know that reducing greenhouse gas emissions is a top priority and that you will be voting for politicians who support effective climate policies. At election time, vote and urge everyone you know to vote as well. Vote in all local, state, and federal elections with global warming being a key determinant regarding who you vote for.

What laws and regulations at the local, state, and national levels do I want to make sure get passed?

_____

_____

_____

_____

_____

Take action in your community

Push for your power company to provide energy from renewables and for electric vehicle charging stations in your area. Push for more public transit and bicycle paths, for better water conservation and waste management, and for local food production and farmers markets. If you live in an ocean-front town or city, push for a study of how rising sea levels will impact your community and what actions need be taken. Make sure your local schools offer programs on climate change.

What community initiatives do you want to push for?

_____

_____

_____

_____

_____

Join the climate movement

As Bill McKibben says, there are powerful ME actions you can take as an individual in your personal life, but what is most needed now are WE actions: collective actions aimed at specific policy outcomes. Find one or more organizations you can join to bring about changes in climate policies and get involved. Attend national and local climate marches and demonstrations. Give money! Financial support is particularly important, so put your money where your heart is. There are literally dozens of organizations you can align yourself with (see the *Resources* section).

What organization(s) am I attracted to joining (and funding)?

_____

_____

_____

_____

_____

Reduce your own carbon footprint

As you learn about global warming, one of the "inconvenient truths" is that if you are living the lifestyle of an average American, you are creating way more than your share of carbon dioxide pollution – more than 20 tons per year. Your own personal carbon footprint reflects all the things you do and consume that adds carbon to the atmosphere. (You could consider your "share" to be the amount that each human could be responsible for emitting while keeping carbon in the atmosphere below the "safe" level of 350 parts per million – it is now over 400 parts per million.) Even people in the United States with the lowest usage – the homeless, Buddhists monks, or sedentary retirees – still produce more than double the global per capita average. If you are going to talk climate action, you will need to look seriously at your own life and lifestyle: are you going to be part of the solution rather than part of the problem? The three main areas contributing to carbon emissions are housing, transportation, and food. There are online calculators you can use to measure your own carbon footprint.

More suggestions for reducing your carbon footprint

Big-ticket actions

Here are the things that could make the biggest difference but are fairly costly – though prices should continually come down.

- Get your home's electricity from solar panels: https://www.nrdc.org/stories/should-you-go-solar.
- Weatherize and retrofit your home completely to become energy efficient.
- Drive an electric car – or at least a hybrid: https://www.nrdc.org/stories/your-guide-going-electric.

Big-commitment actions

There are a few things that won't cost you a lot of money (and may save you some) but have a big impact on your footprint:

- Reduce your automobile driving (bike, take public transit, rideshare, etc.).

- Reduce your airplane flights. (One flight can produce as much carbon as one year of driving.)
- Reduce your overall consumption. (Buy less stuff.)
- Reduce your consumption of red meat, particularly beef.

Small actions that add up

Making energy conservation a part of our daily awareness is essential to the goal of reducing global warming. Here are some things you can do in and around your own home:

- Switch to energy-efficient lighting with CFLs and LEDs.
- Improve the efficiency of home appliances and buy new energy-efficient ones.
- Reduce energy needed for heating. And air conditioning.
- Use less water; shift to solar hot water.
- Reduce lawn size.
- Recycle and reuse whenever possible.
- Eat fresh, locally produced food.
- Eat more vegetarian meals.
- Paint your home a light color if you live in a warm climate, or a dark color in a cold climate. (This can save up to 5,000 pounds of carbon dioxide per year.)
- Choose clean energy options from your local energy provider – request that they provide you with up to 100 percent of your power from wind, solar energy, and other renewable sources.
- Purchase carbon offsets.
- Plant trees.
- Share assets such as cars, houses, tools, etc.

What will I do to reduce my carbon footprint?

_____

_____

_____

_____

_____

Reduce your use of plastics

Plastic refuse on Henderson Island in the South Pacific.

Plastic shopping bags, bottles, and plastic packaging are convenient, but the amount of plastic material that gets into our global waste stream are a scourge on the planet's ecosystem. Ever heard about the Great Pacific Garbage Patch? It is a gyre of trash in the Pacific Ocean estimated to be at least twice the size of Texas that will not dissolve for many decades.

What will I do to reduce my use of plastics?

Reduce the carbon footprint of your workplace

Taking actions to reduce carbon emissions can save money and increase profitability. They may also attract more customers and better employees and investors. There's a ton of information online about business best practices in sustainability. Even as a sole proprietor or a small business you can improve your footprint.

If you work in a big corporation, consider becoming an "intrapreneur" – someone working within the company to make improvements in the company's products or processes that would save energy, water, or other resources.

What could I make happen to reduce the carbon footprint as well as water use at my work/workplace and in my company's entire supply chain?

Your Commitment

1. **The actions I am committing to take:**

   a. **In my personal and community life**

   _____

   _____

   _____

   _____

   b. **In my professional life**

   _____

   _____

   _____

   _____

2. **When I am going to start**

   _____

   _____

   _____

   _____

CHAPTER 6

Resources for Learning and Action

I. Our Major Challenges

    A. Global warming and climate change
        1. Films and videos
        2. Books and articles
        3. Organizations
    B. Related challenges
        1. Overpopulation
        2. Overdevelopment and consumption
        3. Degrading land, water, and ocean resources
        4. Plastics in the environment
        5. Climate refugees
        6. Species extinction

II. Hope for the Future

    A. Future visions
    B. Renewable energy revolution
    C. Economic and business solutions

III. Exploring Values and Consciousness

    A. The evolution of consciousness
    B. Personal growth, spirituality, and encore years
    C. Thought leaders

IV. Environmental Activist and Educational Organizations

    A. Activist organizations
    B. Interfaith organizations
    C. Educational organizations

I. Our Major Challenges

A. Global warming and climate change

1. Films and videos

An Inconvenient Truth
www.algore.com
> Featuring former Vice President Al Gore, this famous documentary film shines a light on the causes and effects of global warming and calls on the human race to address this critical issue while there is still time.

An Inconvenient Sequel: Truth to Power
https://inconvenientsequel.tumblr.com
> A decade after *An Inconvenient Truth* brought climate change into the heart of popular culture, this follow-up shows just how close we are to a real energy revolution.

Before the Flood
www.beforetheflood.com
> This documentary film, narrated by Leonardo DiCaprio and directed by Fisher Stevens, looks at the changes occurring on the planet due to climate change and the actions we can take to bring about a better future.

Earth 2100
www.youtube.com/watch?v=LUWyDWEXH8U
> This documentary film, directed by Rudy Bednar, contemplates what life on Earth might look in the year 2100 if present trends in climate change, population, and resource depletion persist.

Years of Living Dangerously
www.yearsoflivingdangerously.com
> Produced by David Gelber and Joel Bach, and airing on the National Geographic Channel, this Emmy Award–winning

documentary series features first-hand stories from people directly affected by climate change and what is being done to find solutions.

How to Let Go of the World and Love All the Things Climate Can't Change

www.howtoletgomovie.com

Josh Fox, an Academy Award-nominated documentary filmmaker, travels to 12 countries on six continents visiting communities hit by the effects of climate change to explore the human experience of the crisis.

Time to Choose

www.timetochoose.com

Charles Ferguson, an Academy Award-winning documentary filmmaker, looks at the scope of climate change and documents existing solutions and innovators that are working to reverse current trends.

Merchants of Doubt

www.merchantsofdoubt.org
www.sonyclassics.com/merchantsofdoubt

Written by Naomi Oreskes and Erik Conway, this book-turned-documentary reveals the playbook used by industries ranging from tobacco to oil to combat scientific consensus when their products are painted in a bad light.

Chasing Ice

www.chasingice.com

Directed by Jeff Orlowski, this documentary film, which received an Emmy Award for Outstanding Nature Programming, tells the story of the Extreme Ice Survey, a magnificent multi-year record of the Arctic ice captured using time-lapse cameras.

This Changes Everything

www.thischangeseverything.org

Written by Naomi Klein, this book and its companion film, links climate change and capitalism to uncover the basis of environmental destruction, and questions fundamental assumptions about the causes and solutions.

Tomorrow

www.tomorrow-documentary.com
Filmmakers Cyril Dion and Mélanie Laurent offer simple solutions to improve life on Earth and keep it healthy and sustainable. Each of us can play a part, even a small one, and make a difference. Every step counts.

2. Books and articles

Drawdown: The Most Comprehensive Plan Ever Proposed to Reverse Global Warming

Drawdown, edited by Paul Hawken, maps, measures, models, and describes the 100 most substantive solutions to global warming to determine if we can reverse the buildup of atmospheric carbon within thirty years.

Climate of Hope: How Cities, Businesses, and Citizens Can Save the Planet

Michael Bloomberg, former mayor of New York, and Carl Pope, former head of the Sierra Club, team up to propose how we all might solve the climate crisis.

Laudato Si: On Care for Our Common Home

In his second encyclical, Pope Francis draws all Christians into a dialogue with every person on the planet about our common home and what we can do to preserve and celebrate it.

HOT: Living Through the Next Fifty Years on Earth

Part travel journalism and part personal reflection on the future of life for the next generation, HOT, written by Mark Hertsgaard, looks at the challenges humans are facing due to rising temperatures and what we can do in the next 50 years to change our trajectory.

The End of Nature
This famous work by Bill McKibben looks at the impact we're having on the planet and how a philosophical shift in our relationship to the natural world is necessary to make the changes that are needed.

Our Choice: A Plan to Solve the Climate Crisis
In this follow-up to *An Inconvenient Truth*, Al Gore presents the solutions to climate change as being already at hand, and discusses how the real breakthrough needs to come in the form of collective commitment.

Climate Change: What Everyone Needs to Know
An accessible exploration of the science of climate change written by Joseph Romm, a former acting assistant secretary of the U.S. Department of Energy.

Dire Predictions: Understanding Climate Change
Graphical explanations and clear writing from climate scientists Michael E. Mann and Lee R. Kump communicate the scientific basis for understanding climate change.

Eaarth: Making a Life on a Tough New Planet
This look at the current impacts of changing climates from longtime writer and activist Bill McKibben explores the contributing economic and cultural trends, and proposes new approaches to stave off the worst of the possible outcomes.

Field Notes From a Catastrophe
Writer Elizabeth Kolbert provides first-hand accounts of the impacts of climate change, conversations with climate scientists, and a look at corporate lobbying and complicit governments that stand in the way of policy changes.

Don't Even Think About It: Why Our Brains Are Wired to Ignore Climate Change

Written by George Marshall, this book provides insight gleaned from interviews with psychologists, climate scientists, and activists on both sides of the issue. Explores whether humanity is able to accept the research on climate change.

Hot, Flat, and Crowded: Why We Need a Green Revolution — and How It Can Renew America

A proposal by Thomas Friedman to marshall the United States' public and private resources in service of a green revolution.

Down to the Wire: Confronting Climate Collapse

Written by David Orr, a sober and eloquent assessment of climate destabilization and an urgent call to action.

The Ends of the Earth: A Journey to the Frontiers of Anarchy

Author Robert Kaplan travels the world looking at some of the greatest challenges facing humanity: war, population increase, and dwindling resources.

Collapse: How Societies Choose to Fail or Succeed

This famous work by Jared Diamond explores the complex factors that have contributed to the failure of civilizations throughout history.

Up to Five Billion Face 'Entirely New Climate' by 2050

www.climatecentral.org/news/one-billion-people-face-entirely-new-climate-by-2050-study-16587

Written by Andrew Freedman, this article published by Climate Central describes just how drastic changes related to climate change may be in the coming decades.

Short Answers to Hard Questions About Climate Change
https://www.nytimes.com/interactive/2015/11/28/science/
what-is-climate-change.html
> An easy-to-digest article from the *New York Times*, with 16 questions and answers about climate change. Gives an excellent overview of the issue and how to address it.

A World at War: We're under attack from climate change—and our only hope is to mobilize like we did in WWII
www.newrepublic.com/article/135684/
declare-war-climate-change-mobilize-wwii
> In this article in the *New Republic*, Bill McKibben likens our current struggle against climate change to a World War, and argues that we must mobilize in a similar drastic manner if we want to turn current trends around.

The 12 Questions Every Climate Activist Hears and What to Say
https://www.climaterealityproject.org/content/12-questions-
every-climate-activist-hears-and-what-say
> This handy resource offered by the Climate Reality Project lists twelve of the most common arguments against climate change and ways you can respond to them.

Wendell Berry on Climate Change: To Save the Future, Live in the Present
http://www.yesmagazine.org/issues/together-with-earth/
wendell-berry-climate-change-future-present
> Noted poet and farmer Wendell Berry looks at the challenges we face and why we need to live more firmly in the present moment.

The New 50 Simple Things Kids Can Do to Save the Earth
> This kid-friendly guide by Sophie Javna and the Earth-Works Group provides simple tips and tools for how kids can be aware of their carbon footprint and take action early to contribute to the world.

Sustainable World Sourcebook

The Sustainable World Sourcebook is a beautifully illus-trated handbook that provides straightforward solutions and actions for individuals and communities. An essential guidebook for every concerned citizen.

3. *Organizations*

Intergovernmental Panel on Climate Change (IPCC)
www.ipcc.ch

This United Nations body is responsible for assessing the science related to climate change.

United Nations Environment Programme (UNEP)
www.unenvironment.org

The leading global environmental authority that sets the global environmental agenda and serves as an authoritative advocate for the global environment.

National Aeronautics and Space Administration (NASA)
www.climate.nasa.gov

Learn more about the science behind climate change.

Climate Central
www.climatecentral.org

Get up-to-date news and reports on climate change worldwide.

Union of Concerned Scientists
www.ucsusa.org

Explore cutting edge scientific solutions to the most press-ing environmental and societal problems of our time.

B. Related Challenges

1. Overpopulation

"The Earth Is Full"
www.ted.com/talks/paul_gilding_the_earth_is_full
> In this sobering TED Talk, Paul Gilding explains how we've already passed many tipping points for resource use and population on this planet, and how this is a critical time to take action.

Overdevelopment, Overpopulation, Overshoot
https://populationspeakout.org/the-book/
> A collection of powerful images depicting the impact, both environmental and social, of overpopulation and overdevelopment.

World Population Awareness
www.overpopulation.org
> Find articles and recent news related to overpopulation and efforts to address it.

2. Overdevelopment and consumption

The Story of Stuff Project
https://storyofstuff.org
> After the success of her 2007 documentary film, *The Story of Stuff*, Annie Leonard started the Story of Stuff Project as an ongoing effort to raise awareness of the dangers of our consumption-crazed culture.

3. Degrading land, water, and ocean resources

Outgrowing the Earth: The Food Security Challenge in an Age of Falling Water Tables and Rising Temperatures

Author Lester R. Brown explores the growing threat of food shortages and rising food prices, and how addressing these issues must involve moving away from fossil fuels and drastically curbing overpopulation.

Oceans: The Threats to Our Seas and What You Can Do to Turn the Tide

A guide by Jon Bowermaster that explores the health of our oceans, and what we can do to improve it.

The Rising Sea

This book by Orrin H. Pilkey and Rob Young explores the impact of sea level rise on coastal areas and what we can do to address the root causes.

50 Ways to Save the Ocean

David Helvarg's book outlines simple and practical steps one can take to preserve our planet's oceans.

Blue Covenant: The Global Water Crisis and the Coming Battle for the Right to Water

One of the most important books written about the water crisis by Maude Barlow, the head of the Council of Canadians and the Blue Planet Project.

Water: The Epic Struggle for Wealth, Power, and Civilization

This book by Steven Solomon tells the story of the rise and fall of civilizations throughout history using water as the connecting theme between all of them.

Fixing Climate: What Past Climate Changes Reveal About the Current Threat—and How to Counter It

With Wallace S. Broecker as his guide, award-winning science writer Robert Kunzig looks back at Earth's volatile climate history in order to shed light on the challenges ahead.

Scientists predict huge sea level rise even if we limit climate change

www.theguardian.com/environment/2015/jul/10/scientists-predict-huge-sea-level-rise-even-if-we-limit-climate-change

This humbling article by *The Guardian* looks critically at what will likely be unavoidable sea level rise and the impact this will have on coastal regions.

Mission Blue

www.mission-blue.org

An initiative of the Sylvia Earle Alliance (S.E.A.), Mission Blue works to raise public awareness for the protection of key areas critical to the health of our oceans.

4. Plastics in the environment

Plastic: A Toxic Love Story

The story about how plastics has taken over every aspect of our life, by Susan Frienkel.

Plastic-Free: How I Kicked the Plastic Habit and How You Can Too

Written by Beth Terry, how one consumer of plastics awoke to the problems plastics were creating and broke the habit.

5. Climate refugees

Climate Refugees

www.videoproject.com/Climate-Refugees.html

A feature film that explores in-depth the global human impact of climate change and its serious destabilizing effects on international politics. The film turns the distant concept of global warming into a concrete human problem with enormous worldwide consequences.

The Age of Consequences
www.theageofconsequences.com/
Through the lens of national security and global stability, this film looks at the impacts of climate change on increased resource scarcity, migration, and conflict.

American Exodus: Climate Change and the Coming Fight for Survival
Written by Giles Slade, this book looks at how the changing climate may reshape North America.

Climate Wars: What People Will Be Killed For in the 21st Century
Written by Harald Welzer, this book examines the struggles over drinking water, new outbreaks of mass violence, ethnic cleansing, civil wars in the Earth's poorest countries, endless flows of refugees: the new conflicts and forces shaping the world of the 21st century.

Search for Common Ground
www.sfcg.org
For the past thirty-five years, Search for Common Ground has been working tirelessly in numerous countries in Africa, the Middle East, and Asia to end conflicts that often create climate refugees seeking safety. As they note: "Conflict and differences are inevitable. Violence is not."

Millions Face Hunger by 2030 Without 'Deep Transformation' of Agriculture: UN
www.commondreams.org/news/2016/10/17/millions-face-hunger-2030-without-deep-transformation-agriculture-un
This article published by Common Dreams outlines the content of a 2016 UN report that warns that many millions

of people could be forced into poverty due to the effects of climate change.

6. Species extinction

The Sixth Extinction: An Unnatural History
Elizabeth Kolbert reports in this book how we are currently in the midst of a man-made sixth extinction.

The Fate of the Species: Why the Human Race May Cause Its Own Extinction and How We Can Stop It
In this book, Fred Guterl examines many possible scenarios for the future, laying out the existing threats and offering his perspective on the means to avoid them.

Racing Extinction
racingextinction.com/the-film
This film explores global warming, overpopulation, globalization, and animal agriculture as leading causes of humanity's possible extinction.

II. Hope for the Future

A. Future visions

Sacred America, Sacred World: Fulfilling Our Mission in Service to All
A unique and beautiful synthesis of modern politics and spirituality, this book by Stephen Dinan offers the perspective that by bringing these two seemingly disparate worlds together we can create a vibrant future for the United States.

Promise Ahead: A Vision of Hope and Action for Humanity's Future
A powerful counter-narrative to dark predictions about the state of the world, this book by Duane Elgin provides

a compelling blueprint for a possible future that is both hopeful and doable.

Leading from the Emerging Future: From Ego-System to Eco-System Economies
This thought-provoking guide by Otto Scharmer presents proven practices for building a new economy that is more resilient, intentional, inclusive, and aware.

2052: A Global Forecast for the Next 40 Years
www.2052.info
Jorgen Randers draws on the work of more than thirty leading scientists, economists, futurists, and other thinkers to guide us through the future he feels is most likely to emerge.

Journey to the Future
Guy Dauncey's futurist novel considers what a brighter future for humanity might look like in the year 2032.

The More Beautiful World Our Hearts Know Is Possible
This powerful and thought-provoking book by Charles Eisenstein uses individual stories to show that by fully embracing and practicing the principle of interconnectedness—called interbeing—we become more effective agents of change and have a stronger positive influence on the world.

Dreaming the Future: Reimagining Civilization in the Age of Nature
Through a series of short essays, Kenny Ausubel introduces readers to people around the world taking action and shifting thought paradigms to bring about a new future.

Active Hope
Active Hope, by Joanna Macy and Chris Johnstone, shows us how to strengthen our capacity to face the current crises so

that we can respond with unexpected resilience and creative power through processes informed by an intersection of spirituality, psychology, and science.

A Global Vision: General Principles for a Sustainable Planet

This book by Jim Sloman explores the interconnectedness of ecology, energy, finance, geopolitics, and other dimensions of our society, and how our choices now will determine our future.

Ignition: What You Can Do to Fight Global Warming and Spark a Movement

Edited by Jonathan Isham and Sissel Waage, *Ignition* brings together some of the world's finest thinkers and advocates to jump start the ultimate green revolution.

B. Renewable energy revolution

Project Drawdown
www.drawdown.org

Project Drawdown, an initiative of Paul Hawken, is the most comprehensive plan ever envisioned to reverse global warming: the 100 most substantive, existing solutions to address climate change.

The Rocky Mountain Institute (RMI)
www.rmi.org

Cofounded by Amory Lovins, RMI seeks to transform energy use to create a clean, prosperous and secure low-carbon economy.

"The Case for Optimism on Climate Change"
www.ted.com/talks/al_gore_the_case_for_optimism_on_climate_change

This powerful TED Talk by Al Gore examines challenges we're facing regarding climate change and why there's reason to have hope.

State of the World: Is Sustainability Still Possible?
www.library.uniteddiversity.coop/More_Books_and_
Reports/State_of_the_World/State_of_the_World_2013-
Is_Sustainability_Still_Possible.pdf

This publication by the Worldwatch Institute cuts through the ambiguity of the word sustainability and gets to the core of what next steps we need to take together as a species to preserve our planet.

Plan B 4.0: Mobilizing to Save Civilization
www.earth-policy.org/images/uploads/book_files/pb4book.pdf

This book by Lester R. Brown explores our transition to a new energy economy based on renewable energy sources and how it will affect our daily lives.

The Great Transition: Shifting from Fossil Fuels to Solar and Wind Energy

This book by Lester R. Brown discusses the movement away from fossil fuels and towards clean energy sources.

"Optimizing the Energy Revolution"
www.youtube.com/watch?v=PRYYSnZ4I1A

Danny Kennedy, speaking at the Bioneers conference in 2016, outlines how we can double down on our adoption of clean energy and the importance of making this transition on a faster time scale than we have so far.

Fight Global Warming
www.greenpeace.org/usa/global-warming

This article and short video by Greenpeace makes the case for a renewable energy [r]evolution to address climate change.

A Conservative Case for Climate Action
www.nytimes.com/2017/02/08/opinion/a-conservative-case-
for-climate-action.html

This *New York Times* op-ed argues that a carbon tax and dividend system to address climate change is policy that could agree well with both liberal and conservative policy.

Trump, Putin and the Pipeline to Nowhere
www.thenearlynow.com/trump-putin-and-the-pipelines-to-nowhere-742d745ce8fd#.r5s3a1r8n

Author and futurist Alex Steffen describes in this article the concept of a Carbon Bubble and how U.S. economics and politics are being shaped by the impacts of climate change.

Conserve Energy Future (CEF)
www.conserve-energy-future.com

Learn more about diverse types of alternative energy, as well as the causes and effects of different pollutants.

Post Carbon Institute
www.PostCarbon.org

Post Carbon Institute provides individuals and communities with the resources needed to understand and respond to the interrelated ecological, economic, energy, and equity crises of the 21st century.

C. Economic and business solutions

Reinventing Prosperity: Managing Economic Growth to Reduce Unemployment, Inequality and Climate Change
Jørgen Randers and Graeme Maxton make a persuasive economic argument in this book that proves we can all live better lives in this finite world.

The End of Growth: Adapting to Our New Economic Reality
This book by Richard Heinberg posits that the expansionary trajectory of industrial civilization is colliding with non-negotiable natural limits, and explores the resulting impact on our economic systems.

The Real Wealth of Nations: Creating a Caring Economics

This book by Riane Eisler looks at economics from a larger perspective than the powers of the market, arguing that we must give visibility and value to the socially and economically essential work of caring for people and the planet if we are to meet the enormous challenges we are facing.

The Bridge at the Edge of the World: Capitalism, the Environment, and Crossing from Crisis to Sustainability

Author James Speth draws connections between the current environmental crisis and modern capitalism, and suggests that we must change the basic operating structures of our modern economy in order to address environmental degradation and climate change.

Agenda for a New Economy: From Phantom Wealth to Real Wealth

In this book, David Korten describes his vision of the alternative to the corporate Wall Street economy: a Main Street economy based on locally owned, community-oriented "living enterprises" whose successes are measured as much by their positive impact on people and the environment as by their positive balance sheet.

Confessions of a Radical Industrialist

With practical ideas and measurable outcomes that every business can use, Ray C. Anderson shows in this book that profit and sustainability are not mutually exclusive; businesses can improve their bottom lines and do right by the planet.

"Salvation (and Profit) in Greentech"

www.ted.com/talks/john_doerr_sees_salvation_and_profit_in_greentech

Venture capitalist John Doerr gives a moving TED Talk on why he's not hopeful for the future, but sees that investment in green energy might be our only hope.

The Green Collar Economy: How One Solution Can Fix Our Two Biggest Problems

Van Jones, noted TV commentator, activist, and author, makes the case in this book for embracing renewable energy sources in order to address two of the biggest problems facing humanity: environmental degradation and struggling economic systems.

The Necessary Revolution: How Individuals And Organizations Are Working Together to Create a Sustainable World

Peter M. Senge, Bryan Smith, Nina Kruschwitz, Joe Laur, and Sara Schley explore in this book how individual people, in their personal lives and in business, are supporting a powerful shift toward more sustainable ways of living and working.

"Cradle to Cradle Design"

www.ted.com/talks/william_mcdonough_on_cradle_to_cradle_design

Architect and author William McDonough explores in this TED Talk how the design of our products and buildings could change to create a sustainable future for "all children of all species for all time."

The Tactics of Hope: How Social Entrepreneurs Are Changing Our World

Author Wilford H. Welch present twenty-seven case studies of extraordinary social entrepreneurs who have created initiatives to address challenges in health, education, microcredit, fair trade, human rights and social justice, disaster relief, and rehabilitation of the environment.

Intrapreneuring: Why you don't have to leave the corporation to become an entrepreneur

Author Giford Pinchot III shows in this book how intrapreneurs, defined as employees of a corporation, are empowered to explore new ways for the corporation to change its practices and products.

Conscious Capitalism
www.consciouscapitalism.org

Conscious Capitalism supports organizations and individuals to practice conscious capitalism, a way of thinking about capitalism and business that better reflects where we are in the human journey, the state of our world today, and the innate potential of business to make a positive impact on the world.

SustainableBusiness.com
www.sustainablebusiness.com

This website is a hub for sustainability news, networking, and jobs for green business.

Kiva
www.kiva.org

Kiva is an excellent example of a social entrepreneurial NGO. Kiva provides microloans to individuals in 82 countries in need of funds to start their own small business. The funds for these microloans come from individuals with the desire to make a difference in the lives of others more needy than themselves.

American Sustainable Business Council (ASBC)
www.asbcouncil.org

The American Sustainable Business Council is a network of businesses and business associations that have committed themselves to the triple bottom line of People, Planet, and Profit. ASBC members believe that sustainable business is good business, and a sustainable economy is a prosperous and resilient one.

The B Team
www.bteam.org

The B Team is a nonprofit initiative formed by a global group of business leaders to catalyse a better way of doing business, for the wellbeing of people and the planet.

Tim Jackson
www.timjackson.org.uk

Tim Jackson is a British ecological economist and professor of sustainable development at the University of Surrey focused in part on how to create prosperity without growth. He is exploring how our consumerist habits are contributing to the ecological problems we face and what we might do to address them.

Richard Heinberg
www.richardheinberg.com

A senior fellow at the Post Carbon Institute who has written thirteen books on society's energy, economic, and sustainability crises.

III. Exploring Values and Consciousness

A. The evolution of consciousness

Living Deeply: The Science and Art of Transformation in Everyday Life

This book, part of the Living Deeply project offered by the Institute of Noetic Sciences, reveals the perennial wisdom across religions, cultures, and traditions that can help you to live more fully and deeply.

Birth 2012 and Beyond: Humanity's Great Shift to the Age of Conscious Evolution

This book by Barbara Marx Hubbard changes the story that we live in dark times with the idea that the crises we're facing are actually the birth pains of a new world.

The Living Universe: Where Are We? Who Are We? Where Are We Going?

In this book, Duane Elgin sources from the fields of cosmology, biology, physics, and his participation in

NASA-sponsored psychic experiments to show how we are always connected to a living field of existence that makes up reality as we know it.

The Empathic Civilization: The Race to Global Consciousness in a World in Crisis
www.empathiccivilization.com
This book shows the disconnect between our vision for the world and our ability to realize that vision lies in the current state of human consciousness. Author Jeremy Rifkin suggests it will take a restructuring of that consciousness to bring about a better world for all life.

New Consciousness for a New World: How to Thrive in Transitional Times and Participate in the Coming Spiritual Renaissance
Kingsley L. Dennis calls for a paradigm shift in human thinking in recognition of the interconnectedness of all things.

Thrive: The Third Metric to Redefining Success and Creating a Life of Well-Being, Wisdom, and Wonder
In this book, co-founder and former editor-in-chief of the *Huffington Post* Arianna Huffington calls for a reexamination of our modern standards for success.

Reason for Hope: A Spiritual Journey
Noted scientist and environmentalist Jane Goodall explores her beliefs about spirituality and moral evolution in this spiritual autobiography.

Change the Story, Change the Future
Author David Korten believes that the stories we tell ourselves help determine our future and that if we want a different future we must believe that a new story is possible.

Cultivating Peace: Becoming a 21ˢᵗ Century Peace Ambassador

This profound guidebook by James O'Dea provides a holistic approach to peace work, exploring its roots in the cultural, spiritual, and scientific dimensions while providing guidance suitable even for those who have never considered themselves peacebuilders.

Institute of Noetic Sciences

www.noetic.org

The Institute of Noetic Sciences is a nonprofit organization dedicated to supporting individual and collective transformation through consciousness research, transformative learning, and engaging a global community in the realization of our human potential.

The Shift Network

www.theshiftnetwork.com

The Shift Network creates and produces transformational programs, media, and events to help people awaken to their true potential in order to create a bright future for humanity.

Great Transition Stories

www.greattransitionstories.org

Great Transition Stories is based on the insight that the stories we tell shape our experience and that different stories can lead to different outcomes.

New Stories

www.newstories.org

New Stories brings people and projects together around a central flame of nurturing the emergence of new stories for who we are as humanity, what we can become together, and how to navigate the process of change.

Wisdom of the World

www.wisdomoftheworld.com

This media production company uses media to heal, empower, and connect us to the best part of ourselves.

The Earth Charter
www.earthcharter.org/discover/the-earth-charter
An international declaration of values and principles fundamental to creating a just and sustainable future for mankind.

B. Personal growth, spirituality, and encore years

"Less Stuff, More Happiness"
www.ted.com/talks/graham_hill_less_stuff_more_happiness
In this TED Talk, Graham Hill makes the argument that living with less stuff and taking up less space can lead to greater happiness, and outlines three guidelines for making this happen.

Love Is Letting Go of Fear
This classic guide to personal transformation by Gerald G. Jamplosky is designed to help us let go of the past and stay focused on the present as we step confidently toward the future.

Global Spirit
www.cemproductions.org
A documentary series produced and directed by Emmy Award–winner Stephen Olsen that brings to light the various practices – spiritual, mental, and physical – that help us define who we are as human beings.

The Power of Now: A Guide to Spiritual Enlightenment
A beautiful look at the power and opportunity of living in the present moment, this world-renowned book by Eckhart Tolle provides sage advice for how to stay present in the day-to-day challenges of life.

Living in Gratitude: A Journey That Will Change Your Life

This book by Angeles Arrien combines teachings from social science with simple practices and prayers to support people in cultivating a daily practice of living in gratitude.

The Power of Meaning: Crafting a Life That Matters

In this book, Emily Esfahani Smith explores the idea that the search for meaning, as opposed to a search for personal happiness, is what can bring deep fulfillment in life.

Forgiveness: A Time to Love and a Time to Hate

In this two-part documentary film, writer, producer, and director Helen Whitney examines the power of forgiveness through stories that range from the intimately personal to global scales.

The Soul of Money: Reclaiming the Wealth of Our Inner Resources

Author Lynne Twist explores our relationship with money, what it tells us about our values, and how that awareness can add value to our lives and the lives of others.

The Wayfinders: Why Ancient Wisdom Matters in the Modern World

This book by Wade Davis explores how understanding the wisdom of the traditional cultures of the world will be our mission for the next century.

Last Child in the Woods: Saving Our Children From Nature-Deficit Disorder

In this influential work about the staggering divide between children and the outdoors, child advocacy expert Richard Louv directly links the lack of nature in the lives of today's wired generation—he calls it nature-deficit—to some of the most disturbing childhood trends, such as the rises in obesity, attention disorders, and depression.

Planet Earth
A timeless BBC documentary series that explores the wonder and beauty of our amazing planet home.

Finding Meaning in the Second Half of Life: How to Finally, Really Grow Up
In this book, James Hollis explores the ways we can grow and evolve to fully become ourselves when the traditional roles of adulthood aren't fulfilling, revealing a new way of uncovering and embracing our authentic selves.

Composing a Further Life: The Age of Active Wisdom
A collection of stories, this book by Mary Catherine Bateson relates the experiences of men and women who, upon entering their second adulthood, have found new meaning and new ways to contribute, composing their lives in new patterns.

Commonweal
www.commonweal.org

Commonweal works in three core fields—health and healing, art and education, and environment and justice—by supporting specific programs in myriad areas including cancer, health professional education, environmental health, adult learning, yoga, healing nutrition, permaculture gardening, and juvenile justice.

A Network for Grateful Living
www.gratefulness.org

This unique organization offers online and community-based educational programs and practices that inspire and guide a commitment to grateful living, and catalyze the transformative power of personal and societal responsibility.

Attitudinal Healing International
www.ahinternational.org

Attitudinal Healing International works toward a world in which each person takes responsibility for what we think,

say, and do, and where each of us feels safe, whole, and
loved.

Wisdom 2.0
www.wisdom2summit.com
> This perennial global gathering explores the intersection of
> wisdom and technology.

Encore.org
www.encore.org
> This organization, formerly known as Civic Ventures, grew
> out of a desire to transform the aging of America – one of
> the most significant demographic shifts of the 21st century
> – into a powerful, positive source of individual and social
> renewal.

C. Thought leaders

His Holiness the 14th Dalai Lama of Tibet
www.dalailama.com
> A world-renowned spiritual leader and teacher, the Dalai
> Lama preaches living from a place of happiness and compas-
> sion for all things.

Desmond Tutu
www.desmondtutu.org
> Desmond Tutu is a human rights defender and Nobel Prize
> winner from South Africa. He became world famous in the
> 1980s as an opponent of apartheid and has since gone on to
> become a leading voice for social justice and human rights.

Wendell Berry
www.wendellberrybooks.com
> Wendell Berry is a novelist, poet, environmental activist,
> cultural critic, and farmer who uses his gift with words
> and language to raise awareness of our impact on the

environment through modern consumerist lifestyle and unsustainable food systems.

Marianne Williamson
www.marianne.com

Marianne Williamson is an internationally acclaimed spiritual author and lecturer who is also deeply committed to ending poverty and hunger throughout the world.

Charles Eisenstein
www.charleseisenstein.net

Charles Eisenstein is a teacher, speaker, and writer focusing on themes of civilization, consciousness, money, and human cultural evolution.

Bill McKibben
www.billmckibben.com

Noted author, educator, environmentalist, and co-founder of 350.org, Bill McKibben has written more than a dozen books and is an active player in addressing global warming.

IV. Environmental Activist and Educational Organizations

A. Activist organizations

350.org
www.350.org

350.org is building a global grassroots climate movement that can hold leaders accountable to the realities of science and the principles of justice. Their goal is to raise awareness around climate change and reduce CO_2 levels to what is deemed "safe": 350 parts per million.

Greenpeace
www.greenpeace.org
> Greenpeace is an independent campaigning organization that uses peaceful protest and creative communication to expose global environmental problems and promote solutions that are essential to a green and peaceful future.

Idle No More
www.idlenomore.ca
> Idle No More is an ongoing First Nations movement, started in Canada, that seeks to protect indigenous sovereignty and the land and water that are essential to life.

Bioneers
www.bioneers.org
> A celebration of the genius of nature and human ingenuity, Bioneers is a hub of social and scientific innovators with practical and visionary solutions for the world's most pressing environmental and social challenges.

Sierra Club
www.sierraclub.org
> The Sierra Club is one of the largest environmental non-profits in the United States working to protect the Earth's ecosystems and resources through direct action and legislation.

Citizens' Climate Lobby
www.citizensclimatelobby.org
> Citizens' Climate Lobby is a non-profit, non-partisan, grassroots advocacy organization focused on national policies to address climate change.

Rainforest Action Network
www.ran.org
> Campaigns for the forest, their inhabitants, and the natural systems that sustain life by transforming the global marketplace through education, grassroots organizing, and non-violent direct action.

The Global Reality Project
www.climaterealityproject.org

Founded by former Vice President Al Gore, this organization works to bring the world together to solve the climate crisis and make a sustainable future a reality.

Earth Day Network
www.earthday.org

Earth Day Network's mission is to broaden and diversify the environmental movement worldwide and to mobilize it as the most effective vehicle to build a healthy, sustainable environment, address climate change, and protect Earth for future generations.

Slow Food
www.slowfoodusa.org

A global, grassroots organization with supporters in 150 countries around the world who are linking the pleasure of good food with a commitment to their community and the environment.

Global Alliance for the Rights of Nature
www.therightsofnature.org

The Global Alliance for the Rights of Nature is a network of organizations and individuals committed to the universal adoption and implementation of legal systems that recognize, respect, and enforce "Rights of Nature" and to making the idea of Rights of Nature an idea whose time has come.

Earthjustice
www.earthjustice.org/welcome

The nation's original and largest nonprofit environmental law organization, Earthjustice brings expertise and commitment to the fight for justice and to advance the promise of a healthy world for all.

Earth Law Center

www.earthlawcenter.org

Earth Law Center works to transform the law to recognize and protect nature's inherent rights to exist, thrive, and evolve by building a force of advocates for nature's rights at the local, state, national, and international levels.

The Gaia Foundation

www.gaiafoundation.org

The Gaia Foundation works with local and indigenous communities, civil society groups, and social movements to restore a respectful relationship with the Earth.

Generation Waking Up

www.GenerationWakingUp.org

Seeking to ignite a generation of young people to bring forth a thriving, just, and sustainable world.

Global Fund for Women

www.GlobalFundForWomen.org

A global champion for the human rights of women and girls.

Women's Earth Alliance (WEA)

www.WomensEarthAlliance.org

WEA's vision is that the women worldwide who are safeguarding a thriving home are never alone, and that they have what they need—not only to survive but to thrive.

Women's Environment & Development Organization

www.wedo.org

A global women's advocacy organization for a just world that promotes and protects human rights, gender equality, and the integrity of the environment.

The Commission on the Status of Women (CSW)

www.unwomen.org/en/csw

CSW is the principal global intergovernmental body exclusively dedicated to the promotion of gender equality and the empowerment of women.

International Planned Parenthood Federation (IPPF)
www.ippf.org
IPPF leads a locally owned, globally connected civil society movement that provides and enables services and champions sexual and reproductive health and rights for all, especially the underserved.

It's Time Network
www.ItsTimeNetwork.org
An inclusive community of people and organizations working collaboratively to accelerate the full empowerment of women and girls in order to achieve gender equity, evolve democracy, and build fair economies that regenerate the Earth.

B. Interfaith organizations

Alliance of Religions and Conservation (ARC)
www.arcworld.org
A secular body that helps the major religions of the world develop their own environmental programs, based on their own core teachings, beliefs, and practices.

Interfaith Power & Light
www.interfaithpowerandlight.org
The mission of Interfaith Power & Light is to be faithful stewards of Creation by responding to global warming through the promotion of energy conservation, energy efficiency, and renewable energy.

C. Educational organizations

Natural Resources Defense Council (NRDC)
www.nrdc.org
NRDC creates solutions for lasting environmental change, protecting natural resources in the United States and across the globe.

Environmental Defense Fund (EDF)
www.edf.org

Guided by science and economics, EDF's mission is to preserve the natural systems on which all life depends.

Woods Hole Oceanographic Institution
www.whoi.edu

The world's leading independent, non-profit organization dedicated to ocean research, exploration, and education.

Pachamama Alliance
www.pachamama.org

Pachamama Alliance is a global community that offers people the chance to learn, connect, engage, travel, and cherish life for the purpose of creating a sustainable future that works for all.

Climate One
www.climateone.org

Climate One, an initiative of the Commonwealth Club of California, offers a forum for candid discussion among climate scientists, policymakers, activists, and concerned citizens by gathering and distributing inspiring, credible, and compelling information.

The National Outdoor Leadership School (NOLS)
www.nols.edu

A leading outdoor educational school with programs in over fourteen countries.

NatureBridge
www.naturebridge.org

Provides environmental educational programs in six U.S. national parks, including Yosemite, for more than 30,000 children and teens each year.

Roots & Shoots
www.rootsandshoots.org
> A global youth leadership program in nearly 100 countries established by Jane Goodall. Seeks to develop the compassionate young leaders the planet needs.

Partners for Youth Empowerment (PYE)
www.pyeglobal.org
> PYE offers trainings in North America, UK, South Africa, India, Brazil, Greece, and a growing number of countries that empower teachers and students to build supportive, creative communities. Its sister organization, Power of Hope Camp (www.powerofhope.org) operates camps in the U.S. and British Columbia.

Global Footprint Network
www.footprintnetwork.org
> This organization works with governments, companies, and individuals to make data-driven decisions about how to manage resources in a more sustainable way. Use their Footprint Calculator to learn how your lifestyle choices impact the planet and ways you can live more sustainably.

Worldwatch Institute
www.worldwatch.org
> An independent research company dedicated to accelerating the transition to a sustainable world that works for all. Key objectives include universal access to renewable energy and nutritious food, expansion of environmentally sound jobs and development, and addressing population growth through education on healthy and intentional childbearing.

Yale Program on Climate Change Communication
www.climatecommunication.yale.edu
> This program out of Yale University conducts scientific studies on public opinion and behavior in order to inform the decision-makers and promote discussion and education on climate change.

David Brower Center

www.BrowerCenter.org

A nonprofit space for numerous environmental organizations located in downtown Berkeley, California.

The Redford Center

www.RedfordCenter.org

Works to address frontline social and environmental issues through relatable stories, impact films, and campaigns that lead to positive change.

Acknowledgments

I want to acknowledge my editor, Mary Earle Chase, whose deep commitment to this field over many years, as well as her wonderful writing and editing skills, have been of enormous help in the project.

Thanks also to David Hopkins, one of my millennial friends and my partner in researching and writing *The Tactics of Hope: How Social Entrepreneurs Are Changing Our World.* David provided the expertise for publishing and marketing this book with an eye to maximizing the book's impact.

Sharon Donovan was of immense help in formulating communications and promotional strategies designed to maximize the impact of *In Our Hands.* Her commitment to ensuring the sustainability of our planet and humanity has been demonstrated over and over again these past decades.

My thanks also to Ariel Berendt for her stellar work in helping produce the *Resources for Learning and Action* section of the book, which I believe will be a valuable source of inspiration and knowledge for those who want to become part of the solution.

My thanks to granddaughter Kennedy Warner for her work to help me inform colleagues around the world about this initiative.

And finally, great thanks to the book and website team:
Front and back cover design: Ida Fla Sveningsson
Layout and formatting: JETLAUNCH.net
Printing: Selby Marketing Associates
Website design: Goigi
Website automation software: Ontraport

About the Author

Image courtesy of Will Parrinello, Mill Valley Film Group

Wilford Welch has been exploring the driving forces impacting our world for over five decades as a U.S. diplomat in Asia; as an economic development and business consultant in Asia, the Middle East, Europe, and the U.S.; as the publisher of a world affairs magazine that appeared in 26 countries in six language editions; and as the leader of educational trips to numerous countries and cultures.

Wilford developed future planning scenarios for a number of multinational corporations, including Citibank and Toyota. He was the leader of a research team in 1992 that developed *The Wealth of Nations Index* using 63 variables to measure each nation's economic, information technology, and social well-being. In 2008 he was the author of a book entitled *The Tactics of Hope: How Social Entrepreneurs Are Changing Our World.*

Wilford has also been deeply involved in outdoor and environmental education, including being chair of the board of the NatureBridge Golden Gate campus and the National Outdoor Leadership School (NOLS). In 1994 Wilford led the support team on Mount Everest of NOLS instructors, which removed 5,000 pounds of trash off the high camps and introduced the notion of "Leave No Trace" to Everest mountaineering.

Over the past decades, Wilford has created and led several large conferences exploring the issues addressed in this book, including two *Quest for Global Healing* gatherings in Bali, Indonesia, in 2004 and 2007 for over 1,000 participants from forty countries, and the *Beyond Sustainability* conference that took place at the Kilauea volcano in Hawaii in 2010. Wilford has a BA from Yale, a law degree from University of California, Berkeley, and a PMD degree from the Harvard Business School. He lives in Sausalito, California. Visit him at www.WilfordWelch.com and www.InOurHands.Earth.